JONTY CLAYPOLE is a writer, broadcast producer, arts consultant and a person who stutters. He was previously Director of Arts at the BBC and is a patron of STAMMA (the British Stammering Association). In 2021, he was awarded an MBE for services to culture during the Covid-19 pandemic. He currently lives and works between London and Sydney.

wellcome collection

WELLCOME COLLECTION publishes thought-provoking books exploring health and human experience, in partnership with leading independent publisher Profile Books.

WELLCOME COLLECTION is a free museum and library that aims to challenge how we think and feel about health by connecting science, medicine, life and art, through exhibitions, collections, live programming, and more. It is part of Wellcome, a global charitable foundation that supports science to solve urgent health challenges, with a focus on mental health, infectious diseases and climate.

wellcomecollection.org

WORDS FAIL US

In Defence of Disfluency

JONTY CLAYPOLE

PROFILE BOOKS

wellcome
collection

This paperback edition first published in 2022

First published in Great Britain in 2021 by
Profile Books Ltd
29 Cloth Fair
London
ECIA 7JQ
www.profilebooks.com

Published in association with Wellcome Collection

183 Euston Road
London NWI 2BE
www.wellcomecollection.org

1 3 5 7 9 10 8 6 4 2

Typeset in Sabon by MacGuru Ltd
Printed and bound in Great Britain by CPI Group (UK) Ltd, Croydon CRO 4YY

A CIP catalogue record for this book is available from the British Library.

ISBN 978 1 78816 172 5
eISBN 978 1 78283 508 0

For Constance
(Love at first stutter)

Contents

Introduction: The King and I

In the months prior to King George VI's coronation in 1937, a great deal of time, effort and resources went into anticipating, and therefore mitigating, the worrisome issue of his speech. George's impediment wasn't the charming, if somewhat affected, stutter fashionable with the aristocracy, but a juggernaut of blocks, repetitions and slurred sounds.

As a young prince, George had been unable to avoid public speaking engagements, which often left him feeling humiliated. According to friends, his stutter had rendered him 'intensely sensitive', and amounted at times to 'mental torture'.[1] His one consolation was that as the second son of King George V, he was never going to inherit the throne and the relentless speeches and public broadcasts which that entailed. Then his elder brother, King Edward VIII, abdicated a mere eleven months into his reign.

As George's coronation approached, the Archbishop of Canterbury vetoed the suggestion of a live television broadcast, fearing that it might expose the muscular spasms in the King's cheeks and jaw as he struggled to pronounce a particular word. There was, however, the unavoidable issue of the live radio broadcast he would have to make from Buckingham Palace after the ceremony. The Archbishop noticed that although George's speech had greatly improved since he began treatment with his Australian therapist Lionel Logue, it was still far from fluent. He wrote to Lord Dawson, the King's physician, suggesting that they replace Logue with a new therapist, but Dawson dug in, arguing that any change would be merely unsettling.

George was already in a state of immense anxiety, asking rather desperately if the coronation speech could be pre-recorded with his stutter edited out, and then presented as if live. John Reith, the Director-General of the BBC, crushed the idea, suggesting rather archly that the King should decide himself 'whether the deception mattered'. Reith did, however, instruct Robert Wood, the engineer in charge of outside broadcasts, to deploy whatever technological means were available to assist the cause of fluency. Wood practised with the King, showing him how the amplification of the microphone meant he could speak more softly, rolling into words rather than coming at them hard.

In the final days before the coronation, the King, with the help of Logue, Reith and Wood, practised repeatedly. The speech was recorded and played back, and any words that proved problematic were removed. By the day itself, George was well prepared, and the speech – judged on how he read it more than the content – was deemed a triumph, in official circles at least. The press, which at the time was more deferential to royalty than it is today, reinforced this view: the King had become a great public speaker with a warm and strong voice. This remained the official line throughout the following years as war broke out, Britain's towns were ravaged by the Blitz, and the King repeatedly addressed the nation on the need to stand firm. It's the narrative that survives today due, in part, to the Oscar-winning film *The King's Speech* which shows George, supported by Logue, overcoming his impediment to become the people's King. However, thanks to contemporary diarists and sociologists, we know that the general public experienced these speeches rather differently.

'The King broadcast a speech last night which was badly spoken enough, I should have thought, to finish the Royal Family in this country,' wrote the poet Stephen Spender in 1939. 'It was a great mistake. He should never be allowed to say more than twenty words. His voice sounds like a very spasmodic often interrupted tape machine. It produces an effect of colourless monotony.'[2] The diplomat Harold Nicolson wrote that 'it is agony to listen to him – like a typewriter that sticks at every third word'. A more kindly account claims that one speech 'wasn't so bad'[3] but was still 'marred' by his stutter.

A striking example of the anxiety the King's stutter could cause, not only to himself but seemingly the entire nation, is provided by an investigator for Mass-Observation, a social research organisation whose mission was to record the everyday habits, concerns and speech of the British public. Investigations were carried out by mostly anonymous volunteers. An account by one particular woman brings to life a London pub on the evening of VE Day in 1945. While the enduring image of the victory celebrations is that of conga lines snaking round Eros in Piccadilly Circus, the atmosphere was reportedly much more subdued the moment you stepped a street or two away from city centres. While many people had spent the day at thanksgiving services or street parties, the overall mood suggested 'a dumb numbness of relief'[4] rather than joy. After all, the proclamation of victory had long been anticipated, but there was no end in sight to the prosaic realities of rationing and austerity. Lives were irreparably wrecked and all were affected in some way by the abstinence and sacrifices of war.

By nightfall, people were heading home or hunkering down in pubs for the final event of VE Day: a speech from the King. The pub our investigator went to, notebook in hand, was in Chelsea. It was packed full of people, many of whom were drunk, or simply trying to forget their wartime experiences through forced smiles and laughter.[5] The beer kegs were dry and only gin was available. At 9 p.m., the radio was turned on and the room became 'as hushed as a church' with several women at the back leaping to their feet and 'assuming reverent attitudes'. 'There is a sense that people have been waiting all this time for something symbolic,' our investigator wrote, 'and now they have got it.'

'Today we give thanks to Almighty God,' King George began, his voice echoing out through hundreds of thousands of radio sets across the land, 'for a great …' And stopped. A brief moment passed, although it seemed an eternity. A young man in the Chelsea pub giggled. 'Deliverance,' the King said finally and moved on to the next sentence. He was but ten seconds into a thirteen minute address. The pauses and repetitions came thick and fast. The young man, probably drunk, began to impersonate the King's stutter, and became 'the centre of looks of intense malevolence from all corners of the room'. The worst moment came towards the end of the speech. 'Let us turn our thoughts', the King said, 'to this day of just …' He stopped again, tripping repeatedly on a 'T' sound. The investigator noted that 'several women's foreheads pucker and they wear a lacerated look'. 'T – t – t,' went the young man, giggling loudly. The King started the phrase again and had another run at it. 'Of just …' – and it worked – 'triumph and proud sorrow, and then take up our work again, resolved as a people to do nothing unworthy of those who died for us.' He finished his last words, the opening strains of the national anthem were played and everyone, except for a few Marxists, stood up to sing 'God Save the King'. One imagines nobody was more relieved than the King himself. It was inconceivable, as turned out to be the case, that he would ever give a speech of such import again.

George VI's stutter is one of the most widely known and oft-depicted examples of this most common of speech disorders. It highlights many of the mysteries associated with the condition; not least the enduring mystery of what causes it. Biographers have focused on his dysfunctional upbringing and how his rough, unloving father would bellow 'Get it out!' when the young prince got caught on his words. But these stories only explain how his stutter might have been exacerbated, not the root cause. There is the mystery too of how to cure or at least alleviate a stutter. George's relationship with his speech therapist was celebrated at the time, but most of Lionel Logue's techniques for fluent speech now have little currency and

– as contemporary diaries and memoirs reveal – they simply made the King's speech sound strange and hard to follow. Today there is still no undisputed cause or cure for stuttering, although there are theories and techniques which are as popular as Logue's once were.

But the biggest mystery is the one we rarely talk about: why it mattered so much at the time, as it still does, that the King of England should be – or at least appear to be – fluent; that his words should trip effortlessly from his mouth. On quick reflection, the answer might seem obvious. The monarch needs to be a good communicator, able to lead and inspire, particularly during wartime; and that this requires an ability to speak with power and eloquence. These are not qualities we generally associate with stuttering, but by all accounts George was able to express himself with warmth and directness in person. The fact that he stuttered did not interfere with his ability to form relationships, conduct the business of state and hold politicians to account. His problems with communication were exacerbated not so much when he stuttered as when he tried not to.

Imagine for a moment if George VI had broadcast the speeches he wanted to give in his own voice, rather than the one taught him by speech therapists and radio engineers. They would survive today full of stuttering repetitions, but also, I suspect, with the humanity that his friends found in him. They might be easier to follow because, although particular sounds might briefly be blocked or repeated, the cadences would match the meaning of the words he said. Instead, the pressure to hide his stutter meant he spoke through a veil of unnatural modulation.

Lionel Logue taught George to speak in small phrases – what he called 'three-word breaks' – so that he could pause before rolling into the next one. BBC engineer Robert Wood taught George a flattening 'tone formation and lip formation' to help him get his words into the microphone. The text of speeches was scrutinised, with problematic words replaced by synonyms. By the time George opened his mouth, every effort was focused on verbal tricks, which is why he sounds more like a 'tape machine' or 'typewriter' than a human being. His speeches were exercises in simulating fluency rather than hearts-and-minds rhetoric. Even today, they make for uncomfortable listening. Not only because this was a task he could never truly succeed in, but because in striving for fluency his performance was stripped of vigour and meaning.

George's speeches are a reminder that, although fluency and good communication often go together, they are not the same thing. There are those who have no trouble getting their words out, but fail to say anything convincing with them. And there are those who struggle with speech but still connect with a listener. Few would suggest that the scientist Stephen Hawking was a bad communicator, yet he spoke with a synthesised voice. I believe the reason why it was deemed so essential by the British state, the general public, and George himself, that the King should not stutter – even at the expense of his ability to deliver engaging speeches – lies not so much in the act of stuttering, but in the symbolic importance we attach to it.

Stuttering is widely considered a sign of both physiological and psychological dysfunction. According to that view, not only are people who stutter verbally incompetent, but the problem is exacerbated by childhood neuroses and insecurities. When we encounter a person who stutters, we seem to hear inner turmoil translated into sound. If that person is a stranger, it can feel like an unwanted exposure to somebody's deepest weaknesses. Our response, therefore, is often a combination of pity and repulsion, something which has been proven through repeated studies.[6] Stuttering is both a speech disorder and a social stigma.

But the symbolic potency of stuttering extends way beyond the unfavourable light it casts upon an individual. When a person stutters, the usual flow of speech and conversation is ruptured. This is challenging for us because we think of speech as the oil in the machinery of human society. Words enable us to communicate ideas, to decree laws, to share the secrets of the heart. Although email and social media have further empowered the written word as communication – much as letter writing did in the past – we still prioritise the spoken word for the things that matter. Only when our governments have debated and amended proposals do they become preserved in written law; political briefings describe what a public figure will say in an hour or a day's time, but remain speculative in the eyes of the media until the words have emerged from that person's mouth. In our private lives, we consider speech the appropriate mechanism for both the most joyous and confrontational moments of our lives. 'I love you' should be spoken first, and God forbid the person who tries to end a relationship by text message. A speech impediment like stuttering is more than a verbal handicap then, but an unwelcome blockage in the flow of life.

These negative perceptions are often unconscious, but they determine how we encounter people who stutter on the rare occasions that they are thrust upon us. In the case of George VI, this was magnified into a national concern. As King of the United Kingdom, the Dominions of the British Commonwealth and Emperor of India, George VI was the embodiment of the state. For him to be flawed in person suggested the state itself might be flawed – metaphorically at least. The fluency, or rather disfluency, of the King's speech became as symbolic of the nation's fate as the ravens in the Tower of London. No wonder the anxious, 'lacerated looks' across the country that accompanied every pause in his speeches, and the surges of relief when he pushed through to the next word.

George VI is one, admittedly exceptional, example of a person who stutters, just as stuttering itself is just one of many conditions that affect the human voice. Today, there are many who have what are called speech 'disorders' and spend their lives in fear of their own voices because of the accompanying stigma. We know such disorders exist, but what is rarely appreciated is just how widespread they are.

Statistically, well over a million people in the UK are deemed pathologically non-fluent in verbal speech. The Royal College of Speech and Language Therapists puts the number of children who have some form of speech and language impairment as high as 9 per cent and claims that 20 per cent of the population has some form of communication difficulty at some point in their lives.[7] They may have a stutter; the linguistic impairment of aphasia following a stroke; the coprolalia (swearing) of Tourette's syndrome; the dysarthria of speech associated with cerebral palsy and Parkinson's disease or the distorted voices of dysphonia. All conditions that would have sent the Archbishop of Canterbury in George VI's day into a spin.

The sheer number means that all of us in some way live with a diagnosable disorder, either because we have a condition ourselves or because someone in our family and wider social network does. How we engage with those who have a disorder – whether as a parent, friend or colleague – can dramatically affect their quality of life. Yet these are conditions we often have little understanding of, even when they disturb our own speech. One reason for this is that fluency is considered not just normal, but a necessity. Speech disorders are to be hidden rather than understood. Unfortunately, this all too often leads to exclusion and discrimination because they are, for the most part, not conditions which go away.

This is something I have a personal stake in, both as a person who stutters but also as part of an extended family that does so. That there are several of us is not unusual, for there is strong evidence that stuttering is a hereditary condition. My mother passed it on to me, just as my wife inherited it from her father. In fact, my struggles with speech and language go beyond stuttering. I had developmental delay in my speech as a child, and in my teens was diagnosed with both cluttering and dyslexia. But stuttering is the condition that defined me most. I am unable to separate it from my identity because it was always there, submerged but palpable in the murky half-memories of my childhood, and still present in the hidden blocks and word substitutions of my apparently fluent speech today.

When I enquire of my mother when it started, she

tells me she knew something was awry not long after my second birthday. While most infants chatter adoringly at this stage, stringing together short sentences, I was stubbornly mute. Eventually, she took me to see the famous paediatrician Hugh Jolly, who bounced me around a little, looked inside my mouth, and concluded that I was merely lazy. When I did finally deign to speak, coming on for three years old, a whole sentence fell out and they kept coming. But the words began to get stuck too. I have early memories of being stunned as relatively simple words like 'where' or 'when' disintegrated into a string of bizarre wah-wah sounds; the looks of mild concern on the faces of adults; and, inevitably, the look of indulgent malevolence on my sister's face at the discovery that her imposter younger brother was so wonderfully malfunctioning.

My awareness of having a problem was at first far greater than any trouble it actually caused me. It was something I experienced through the reactions of other people more than the mild inconvenience of sometimes not being able to get words out. I spent fifteen years in and out of speech therapy: from one-on-one sessions with a speech-language pathologist to a two-week bootcamp for chronic sufferers at the Michael Palin Centre for Stammering Children. My mother was a features writer and so my progress was tracked through a string of articles for *Good Housekeeping* and the *Guardian*, generally accompanied by a picture of me with bowl haircut, toothy grin and NHS specs. 'How My Son Lost the Edge of My Wretched Tongue', ran one headline in February 1991 with the byline: 'Anne Woodham on her family's battle against a stuttering blight'.

My anxiety about stuttering grew until it became a defining feature of my life. I put a great deal of energy into avoiding words or situations which might expose me. I was so scared of being bullied I would sometimes pretend to forget my own name when called upon to say it aloud in class. On my first day at university, I slipped a letter under my professor's door begging to be excused any group reading or recital exercises that might be required while completing a degree in English Literature. I so desperately wanted university to be an end rather than an extension of the humiliations of school.

At the same time, only those I was closest to knew my secret for I had developed a rapid-fire way of speaking, pivoting around difficult words or phrases and substituting or paraphrasing with others in a manner which passed for quick-wittedness. After leaving university, I was drawn to broadcast, film and theatre, but carefully built a career for myself behind the scenes. The people I work with are often hyper-fluent: breathtakingly articulate, brilliant performers, able to speak off-the-cuff on myriad subjects. I wonder if, without even knowing it, I became a linguistic groupie, hovering around people who speak in a way I never dared believe I might be able to imitate.

Over the years, my stutter and I reached an uneasy truce: as long as I didn't cross agreed parameters, it would leave me more or less alone. But in my early thirties it inexplicably worsened again. A speech therapy course lasting several months at London's City Lit Adult Education College resulted in a level of fluency I had never known before. Having some emotional distance was a relief, but I also found myself floundering. After so much time digging at the roots of my speech, I didn't know what to focus on and so I tried to make sense of everything I had gone through.

Ever since the age of five or six, I had been locked in a cycle of mitigating tactics to try and conceal my stutter, while discreetly and with a few trusted elders exploring what might be causing it and how it might be cured. Most books on the subject are what I consider cause-and-cure self-help manuals that look at a condition not only in isolation from others, but from fluent speech in general. I saw stuttering, incorrectly, as my unique stigma and could count on one hand the number of times I had encountered it outside of speech therapy. But increasingly I was aware of how prevalent it is. Many people who stutter are closeted or 'interiorised', going to great lengths to hide their condition because they believe it will prove detrimental to their careers as well as their social and romantic lives. And I became aware of other people who in their own way, whether with diagnosable conditions or in an uncategorised void, were struggling with speech. Looking back, it astonishes me how my most worrisome preoccupation was also the one I was most ignorant about. In my desperation to cure or hide it, I never had the courage to bring it into the light and decide for myself what it means to stutter. Ten years ago, I set out to do just this.

This is a book about what happens when speech breaks down, and why we are so afraid of it. It is about the disfluencies that impact all of our lives and their relationship with the prized but elusive state of fluency we hold so dear. It is both a scientific and cultural study. It has to be. While it is now widely accepted that most speech disorders are neurological in origin, they are highly attuned to social and cultural context. In some cases, as with stuttering and the vocal tics of Tourette's syndrome, they intensify or alleviate depending on where the individual is and who else is in the room. Other conditions like aphasia and dysarthria are diagnosed not only according to the symptoms of the subject, but by how intelligible that person is deemed by society. For this reason, I have not attempted a definitive survey but focused instead on the psychological and cultural significance of speech disorders as a whole, focusing on those I am most familiar with, rather than anatomising them all.

The sheer prevalence of speech disorders makes them a concern for everyone, but so do the wider truths they reveal about how we all experience language. While many of us have a diagnosable condition at some point in our lives, there are plenty of others who feel some degree of inadequacy about their speech. Not only do we often struggle to find the right words, but – amid the umming and ahing – sometimes struggle to get them out too. The fear of public speaking is in part a fear of falling on the wrong side of some invisible divide separating ourselves from a charismatic and fluent elite; no wonder it is one of the most widespread of recurring nightmares. Those with diagnosable speech conditions sit on the front line of these anxieties. The intensity of their experiences can provide invaluable insights not only into how the rest of us feel, but into language itself: the hold it has over us, the way it perplexes and torments as much as it clarifies and connects.

I believe that rather than merely tolerating speech

disorders, we need to celebrate them because of the diversity and innovation they bring to human thought and language. Despite first appearances, those who struggle with speech tend, out of necessity, to be linguistic virtuosos. People who stutter develop vast vocabularies of synonyms to replace troublesome words. Those with articulation difficulties choose their words more carefully, for each one is a precious commodity requiring more deliberation than the rest of us put into entire sentences. In finding themselves lost for certain words, people with aphasia learn to contort new meanings and combinations out of those they can recall. For the individual with vocal tics, this linguistic virtuosity is often beyond control, but no less remarkable for it.

Living with these disorders requires summoning a level of linguistic creativity each day that most of us experience only occasionally, if ever. In the right hands, such creativity becomes art. A strikingly disproportionate number of our greatest artists were and are people who have pathologically struggled with speech. Writers like Lewis Carroll, Henry James, Elizabeth Bowen and Christy Brown; philosophers Ludwig Wittgenstein and Stephen Hawking; actors Samuel L. Jackson, Nicole Kidman and Marilyn Monroe; songwriters Edwyn Collins, Kendrick Lamar, Carly Simon and Bill Withers; and political orators and visionaries Aneurin Bevan and Greta Thunberg. All of them are people who have a remarkable ability with language which I believe must, on some level, be connected to their struggles with it.

Through charting the stories of such individuals, we learn not only to appreciate speech disorders but also to

look differently at the unquestioning reverence we have
for fluency in our culture. Although many effective speakers are blessed with a 'smooth tongue' or 'the gift of the
gab', this is not essential for good communication and can
even be antithetical to it. Fluency deconstructed is often
little more than a string of interconnecting clichés and
figures of speech with relatively little content, that may
even conceal untruths and manipulative lies – much like
the 'fake news' we hear so much about today. Sociological
studies continually show that much, or even most, communication between humans is non-verbal: gestures, tone
of voice and body language.[8] Our speech is often little
more than a cover or pretext for the far more primal yet
elaborate exchanges going on beneath. This doesn't mean
that language isn't important to us – it is, particularly
when trying to articulate abstract emotions or ideas – but
the way we use it has far more to do with conforming to
social expectations than translating our deepest and most
complicated thoughts into words.

In the first half of this book, I provide an overview
of some of the most common speech disorders: not
only their physical symptoms, but also the psychological
impact they have on an individual, and the stigma they
carry. (This is not as simple as it sounds as definitions
and diagnoses continue to vary.) The point is to show how
much those with such conditions have in common, and
the benefits that come from thinking about speech disorders as a group.

The following chapters describe the role of culture in
determining our attitudes to both disfluency and fluency.
I introduce the notion of 'hyper-fluency' to describe the

expectations that twenty-first-century communications and labour markets – from social media to gig economics – place on us to appear fluent at all times. Through emphasising the cultural perception of speech disorders, I aim to show that there is nothing objectively 'bad' about them; we have just learned to see them as deficiencies. Nothing reveals this more than the strange and barbaric story of their diagnosis and treatment over the last 150 years, which I follow with a cautionary reminder about how little we still know today. From there, I abandon the search for a cause and cure of speech disorders as (mostly) futile and propose a different way of thinking about them.

Neurodiversity is a concept which emerged in the late 1990s and looks at a wide range of conditions not as diseases but simply as forms of human difference that have unrecognised positive qualities as well as the much documented negative ones. It is rarely applied to neurological disorders of speech, however; an oversight which the second half of the book tries to address. I argue that speech disorders embody a set of traits that often result in better rather than worse communication, greater rather than diminished productivity, and play an important role in challenging some of the prejudices and errors of language as experienced by a fluent mainstream. Neurodiversity is significant because it presents the best opportunity to date of improving the experience of the many millions alive today who struggle pathologically with speech.

Mindful that many with speech disorders struggle to be heard – both metaphorically and literally – I refer to written accounts or the dozens of interviews I have conducted wherever possible. While some of the arguments that follow diverge from the dominant views of speech-language therapy, I hope none of them detract from a profession and body of knowledge that has ultimately had more impact in diagnosing and alleviating the suffering of those with speech disorders than medical science has. It is rarely acknowledged that the practice of listening to patients – not only literally, in terms of monitoring their speech, but also responding to their emotional and psychological state of mind – owes more to speech-language therapy as a discipline than psychotherapy, although it is rarely credited for it.

Finally, a note on terms: while there are those who increasingly talk about speech 'differences', I continue to use 'disorder'. This is because it will be more familiar to most readers and because it may well be retained and rehabilitated by those with such conditions in the same way that the term 'disabled' has. While everybody experiences some 'disfluency' of speech, the term here mostly refers to those pathological conditions where such interruptions become compulsive, problematic and ultimately diagnosable for an individual. I distinguish therefore between 'a disfluency' and 'everyday disfluencies'. In keeping with current practice, I talk about 'people with' stutters, aphasia and tics rather than 'stutterers', 'aphasics' and 'tourettics' – it may be cumbersome, but it is a small price for acknowledging that a speech disorder is a trait rather than the essence of a human being.

I began writing this book partly to alleviate the shame and fear I had about stuttering, but also as a call-to-arms for those who have experienced speech disorders to speak out – in their own voices – about the creativity and productivity, as well as the lows and frustrations, of disfluency. Only this way can we address what most needs to change: not an individual's speech, but the way society thinks about it.

In all of what follows, I have one very simple argument to make: what are generally referred to as 'speech disorders' are simply forms of vocal and linguistic diversity. Despite appalling stigmatisation, they still manage to enrich our language, our ideas and art forms. They forge stronger and more meaningful forms of communication between human beings. Just think what might be possible if we stop resisting and embrace them.

1

Maladies of Speech

It is the late 1940s and Kenya is under British rule. In a family compound some twenty miles outside of Nairobi, A young boy who will become known to the world as Ngũgĩ wa Thiong'o grows up among his four mothers and many siblings. At home he speaks Gikuyu, a Bantu language spoken by barely a quarter of the Kenyan population. But when he passes through the school gates, he is required to speak the language of the colonial authorities. Children caught speaking Gikuyu are often compelled to wear a placard saying 'I am stupid' around their necks.

Already Ngũgĩ is aware that speech is far more than an enabler of communication; it is also subject to dangerous prejudices and manipulations. He knows that some ways of speaking are considered superior to others, and that the enforcement of the English language is central to the subjugation of the Gikuyu people. It is enlisted in what he will later describe as 'mental colonisation'. Alongside such weaponisation of speech, he knows something else too: that the human voice often goes awry in unpredictable ways beyond anyone's comprehension.

'I grew up with stuttering in the family,' Ngũgĩ tells me, out of the blue. We are sitting in a rented apartment overlooking the Grand Canal in Venice on a wet, winter's day; a far cry both in time and place from his childhood home. As Kenya's most famous living writer, his life is a semi-nomadic one of lecture tours and residencies. Finding the opportunity to meet him is a question of getting a ticket to whatever city he happens to be in. For someone like me who is fascinated by the way language is both used and abused the trip is worth it.

Ngũgĩ has thought about this subject, and borne the brunt of its realities, as much as anyone alive. In the late 1970s, he even spent a year in prison, illegally detained by the Kenyan government, because of his decision to write a play in his native tongue, something perceived as a threat to a regime that promoted English and Swahili as its proper languages. There is something in the way Ngũgĩ has repeatedly described the marginalisation of minority languages that resonates for me with the way speech disorders are treated in our society. We have talked for almost two hours, but this intimate recollection of stuttering in the family compound comes as a surprise.

'I had four mothers and one father,' he continues. 'The senior mother had two kids, older than me. My brother Gitogo had a defect of speech. He could utter one word occasionally, but he always struggled to get the word out so mostly he spoke through gestures.' As a consequence, Gitogo wasn't deemed much use for anything at all. But Ngũgĩ loved him and was devastated when Gitogo was shot in the back by British soldiers who assumed, incorrectly, that this strange young man was a militant in the Mau Mau independence movement. Ngũgĩ was still a boy and the murder of Gitogo shaped him indelibly. Although he himself did not have a speech disorder, there was something in Gitogo's struggle with a language that was, in any case, marginalised, that made Ngũgĩ fascinated in the way speech functions and dysfunctions in human society.

Such a fate – face down in the dust, his life bleeding out of him – could easily have come to Ngũgĩ, as it did to so many other young Kenyans, but he was saved by his talent. While Gitogo had struggled to speak at all, Ngũgĩ mastered not only his people's language but the tongue of his masters too. This was critical: proficiency in any academic subject came secondary to the imperative of language. You could be a genius at maths or science, but without fluent English higher education was out of reach. Ngũgĩ's linguistic dexterity earned him a place at Makerere University in Kampala. His first English-language novel, *Weep Not Child* (1964), published when he was only twenty-six, brought him fame and the opportunity to escape the bloodshed at home.

Ngũgĩ never forgot the endless humiliations of his childhood. Their shadow and the fate of his brother Gitogo hang over his writing, in which the experience of inarticulation and repression are always combined. For fifty years his novels charted the changing fortunes of Kenya under British rule, the early hopefulness of independence, and the oppressive dictatorships that followed, but his characters – always pawns at the hands of indomitable forces – struggle to get their words out.[1]

Wizard of the Crow (2006), his late career masterpiece, is set in a fictional version of modern Kenya, rife with corruption and personal insecurity. Slowly a disorder takes over the land in which people are unable to speak.

It begins when they try to articulate their deepest desire with the words 'if' and 'if only'. The wizard of the book's title, a sort of accidental witch-doctor, gives his prognosis: 'There is a strange illness in the land. It is a malady of words; thoughts get stuck inside a person. You have seen stutterers, haven't you? Their stammer is a result of a sudden surge of thoughts, or calculations, or worry.' This malady of words proves contagious, affecting many hundreds of thousands including the corrupt President himself, and only alleviates when society itself begins to change for the better.

Ngũgĩ's image of a malady of words seems fantastical at first, but is less so when one remembers the overwhelming number of people who have speech disorders in real life. It reminds us too that despite the different diagnoses and complicated terms we use, they all serve to exclude such individuals from whole areas of human discourse; not only the affairs of the state and public life, but those of the heart too. And it reminds us that such disorders will only be 'cured' not when the afflicted individual changes, but when our discriminating society is willing to accommodate them.

All these are important observations to keep in mind as we chart the ways speech disorders are experienced in our society today. As is the central theme of Ngũgĩ's life and work: speech, whether 'fluent' or 'disfluent', is never just about communication, but power and subjugation. It is a source of prejudice and inequality. What is at stake goes far beyond day-to-day humiliation; it determines the potential of an individual's life.

Although statistics tell us that hundreds of millions

across the world experience a diagnosable speech disorder at some point in their lives, it is not immediately clear how many such conditions there are, nor indeed what separates 'disordered' from 'normal' speech. Addressing these questions is a necessary first step in any enquiry into the ways in which speech breaks down. After all, most of us encounter some problems with our speech, whether the perennial challenges of turning abstract thought into spoken word, or the passing hesitations, stumbles, malapropisms and blockages that are present in even the most fluent speaker.

These are usually mere hiccups. We are hardly aware of them at all, or, when they do pass through our consciousness, it may be as a passing irritant rather than a recurring problem. But then there are those for whom the process of speaking the words they have in mind is a consistent battle. Since their intelligence is unimpaired, the consciousness of a recurrent breakdown between thought and speech is intensely frustrating. It can feel like taking a step forwards only to find your leg doesn't move. And it is in such consistently bumpy or obstructed linguistic territory that we may identify a disorder.

There are many different types of speech disorder, but all of them share a single quality: the consistent and obtrusive struggle in communicating the words you want to say. Whatever the cause, whatever the symptom, the emotional impact on a life can be devastating. Often speech disorders are experienced alongside, or in consequence of, other conditions like cerebral palsy, stroke or motor neurone disease. One of the 'purest' in this regard, while also the most widespread and enigmatic, is my own condition: stuttering.

It is said that five per cent of children stutter at some point, while one per cent of the adult population continues to do so.[2] If that sounds small, we need only consider that it amounts to nearly 700,000 people in the United Kingdom, over three million in the United States and seventy million globally. In most cases we do not know the cause, although it is increasingly considered a neurological condition, something to do with the wiring of the brain, that affects speech. While it seems simple enough to those who don't experience it – it's the jack-hammering speech of popular caricatures like Porky Pig or Ronnie Barker in *Open All Hours* – for those who do, it covers a vast range of vocal dysfunctions.

Although the parameters of the term have changed, with different types of non-fluency creeping in and out, one well-regarded definition is that of American speech therapist Marcel Wingate. Stuttering, he wrote, is a 'disruption in the fluency of verbal expression, which is characterised by involuntary, audible or silent, repetitions or prolongations in the utterance of short speech elements ... These disruptions usually occur frequently or are marked in character and not readily controllable.'[3] This is a specific, unified definition, but there are also those experts, like linguist David Crystal, who identify 'stuttering' as a catch-all word to cover several kinds of non-fluency, each of which can vary considerably from speaker to speaker.[4]

Although the dominant characteristic is a physical struggle to get words out, no two people experience this in quite the same way. The most widely recognised symptom is the unnatural repetition of sounds, which writers tend to homogenise into a c-c-c-cliché of written form. Then there's the abnormal prolongation of sounds that seem to get trapped in the speaker's mouth. And then the silent block, where no sound comes out at all. There is no universal pattern for sounds more likely to cause obstruction. For some, plosives (like *p*, *d*, *g* and *b*) are particularly problematic, others struggle with fricatives (like *f* and *th*) or elongate vowels. Many experience stuttering as a pot pourri of all three. I have met one person who fears 'Rs and Ls', another who dreads 'K' and 'T' sounds, and another for whom it is 'M', 'W' and 'L'.

While one speaker's experience of stuttering is different from another's, an individual's own symptoms are equally inconsistent. From moment to moment, day to day, behaviour can vary wildly. People who stutter are frequently taken by surprise and if they sometimes look startled in the act it is because they are. Even the most reliable of sounds can prove false friends. This variability is reflected in the range of terms we have for it. Although some people insist on a distinction, 'stuttering' is entirely interchangeable with the word 'stammering'. Americans tend to use the former; British the latter (I use 'stutter' simply out of choice, because its repeated 'Ts' seem more suggestive). There are other words too: 'psellism' and 'dysphemia' are both medical terms; and there are obsolete Shakespearean words like 'mammering' and the prim Victorian euphemism of 'hesitancy'.

These days a distinction is drawn between stuttering and 'cluttering'. Rather than involving excessive breaks in speech, cluttering results from disorganised speech planning. Individuals speak too fast or in spurts which collapse into incomprehension. When I was a child, the therapists at the Michael Palin Centre for Stammering Children measured both the speed and flow of my speech and concluded I cluttered as well as stuttered. My speech would race along then fold into itself leaving me babbling and gasping. I experienced both stuttering and cluttering as different sides of the same disorder and found it hard to distinguish between them except by the sounds they made.

Quite understandably, and at the risk of tautology, there is a tendency to think of people who stutter as people who stutter. But many adults like me had traumatic experiences at school and our disorder became covert or 'interiorised'. You will probably know some of us without being aware of it. We can speak in a slow, incredibly considered and therefore slightly infuriating way, because we are using a great deal of energy and mindfulness to avoid stuttering. Or we might create long pauses at the wrong moments in our speech; not after making some clever or amusing point but simply mid-sentence or even mid-word. We may appear rather inarticulate, choosing unusual phrases and long, circumlocutory sentences to make quite simple points. I consider these to be vocal symptoms of stuttering just as much as the stutter itself.

For some, a stutter is a disability, narrowing access to vast areas of human experience – whether jobs, relationships or vocational fulfilment. For many others, like me, it is a concern but a manageable one that can be mitigated and even entirely concealed through word substitution and voice modification techniques. Then there are those who rarely get through a sentence without getting stuck, but are fluent when speaking a foreign language. When Cardinal Villeneuve encountered George VI in Quebec, he was struck by the way the King stuttered when delivering his speech in English, but was fluent when he repeated it in French.[5] Sometimes a mere change in accent can be enough. When I spoke to the Irish novelist Colm Tóibín about his stutter, he recalled knowing 'a man in rural Ireland who spoke with a posh accent because when he spoke posh he didn't stammer'.

Many find they can speak without a hitch if on their own at home; the problem arises once another person is involved. And there are actors and singers who stutter off-stage but become fluency itself when performing: Rowan Atkinson, Bruce Willis and Emily Blunt are famous examples. It is said Marilyn Monroe developed her distinctive whispery voice not as a technique for seduction but just for getting the words out, and built a career out of it. One theory is that when performing we use different parts of the brain, particularly our memory faculties, compared to when we're engaged in spontaneous conversation. In other words, if stuttering really is something to do with the wiring in the brain, then performance may rewire it – if only for a short duration.

This waxing and waning is one of the most perplexing aspects of stuttering. After all, many disabilities aren't quite so context-dependent, disappearing or re-appearing according to the type of activity the speaker is participating in. Through the Michael Palin Centre for Stammering Children, whose courses I attended as a child, I met British actor Oliver Dimsdale, who has appeared in such popular shows as *Father Brown* and *Downton Abbey*. Oliver is one of those who stammers in person but never on stage or on screen. He tells me how debilitating his stammer was ('cripplingly bad') throughout childhood until, at the age of thirteen, he discovered acting.

In conversation, Oliver's stutter is discernible, but within seconds he can conceal it. 'I sit down,' he says, suddenly speaking slowly, evenly and with regular pauses. 'I get to the core of my breath. I'm looking around the room. I look people in the eyes. I'm taking my time. And when I have an impulse to say whatever it is I have to say next, I could probably talk for hours and hours and not stammer.'

This self-conscious manner of speaking is similar to the process of rehearsing and ultimately performing a part; a process in which Oliver's speech becomes more fluent as the role develops. 'When you know exactly the sentences, the breath and the moves you're going to make – and you're in it, going from moment to moment – you have the freedom to forget you are a person who stammers. You are you, imagining yourself inside the words, inside the head, of another person. I find the fluency inside me for whatever character I'm playing.' I ask him why he doesn't do this the whole time. 'Because it's a bit more exhausting, actually,' he says, his stammer suddenly reappearing. 'There's so many layers of stress that come with auditioning, presentations and teaching, so I'd much rather feel that I was being completely myself in my day-to-day life rather than trying to come across as someone who is fluent all the time, even to my wife and kids.'

Another strange characteristic of stuttering is its gender

bias. Characters in films and popular media who stutter are almost always male, and there's good enough reason for this: throughout the world, there are around four or five times as many men as women who stutter. Nobody knows why. Carolyn Cheasman runs the esteemed speech therapy department at City Lit in London and is one of the leading therapists in the UK. She is also a person who stutters. When I ask her to explain this gender imbalance, she insists there are no facts, only theories, but they are intriguing nonetheless.

One theory draws a link with the way infant boys tend to lag behind girls in speech development. So while girls are better able to express their thoughts with words, boys struggle to do so. Since stuttering tends to emerge between the ages of three and six, at precisely the moment when the developmental gap between boys and girls is at its greatest, there could be some connection between the two phenomena; as if stuttering is caused in part by a lag between a child's conception of relatively sophisticated thoughts and feelings, and their ability to express them.

But Cheasman has another theory, based on her own experience, which is that females might be better, or more determined, to hide a stutter. One of the courses she runs, which I attended in my early thirties, focuses on interiorised stammering for those individuals who construct their speech and lives around concealing their impediment. More women take this course than any of the others. Is it possible that because of certain societal pressures – the expectation to be a certain way – women may be going the extra distance to conceal their struggle with speech? Men, on the other hand, whether for physiological or cultural reasons, seem less willing or less able to conceal it.

People with interiorised stutters can often pass as fluent, although we sometimes spot the tell-tale characteristics of word substitution or speech modification in one another. Sarah is an arts administrator in Manchester who I met through work. One evening, protected by the roar and noise of an industry dinner, we began to swap notes. Like me, she is in her forties and still constructs her professional life around concealing her stutter; avoiding not only words, but certain types of meeting or presentation. I ask her in what ways her experience as a woman might be different from that of men. To my surprise, she has found that stuttering is linked to her menstrual cycle. She can pinpoint almost to the day when her speech is likely to be most fluent or disfluent. As a result, she has long been in the habit of consulting her cycle before scheduling public speaking engagements.

With so many variables and unknowns, stuttering remains a mystery. Uri Schneider, a prominent American speech therapist who runs a large practice treating different disorders in several cities around the world, describes it to me as 'the most enigmatic of speech conditions. If somebody has cerebral palsy and dyspraxia,' he says, 'they don't wake up one day with one type of speech and another day with a different kind of speech. It's a consistent issue. Same with learning disabilities. But stuttering has that erratic, unpredictable quality to it.'

Stuttering is an enigma then. Always has been and perhaps always will be. But I would argue that there is one condition, which, while not in itself a speech disorder, often results in disordered speech and is scarcely less enigmatic or unpredictable than stuttering. It takes its name from the French physician Gilles de la Tourette who first argued in the 1880s that in certain cases motor and vocal tics were not signs of some other condition, but amounted to one in its own right: *La maladie de tics de Gilles de La Tourette,* as it became known.

These tics include involuntary movement of different parts of the body as well as uncontrollable sounds and speech. One of the speech disorders we associate with Tourette's is coprolalia, or the involuntary and repetitive use of obscene language. Although it affects only about one in ten people who have Tourette's,[6] it has become a defining feature because of media and public fascination. Other Tourettic speech disorders include echolalia, which is the repetition of one's own or others' words or phrases; and palilalia, in which a person repeats their own words. Many others with Tourette's find their words are unaffected, but they are interspersed with barks, grunts, yelps and coughs. Once considered extremely rare, it is now thought that Tourette's syndrome – with or without its accompanying speech disorders – affects about one per cent of the population at some point in their lives.[7]

Jess Thom is a performing artist with Tourette's with whom I have collaborated on a couple of television programmes over the past few years. Like stuttering, Tourette's often begins in childhood and alleviates as people grow older, but it is more part of Jess's life in her thirties than ever. Because her motor tics affect her limbs, she uses a wheelchair and other aids to help her through the day. Since people who tic are often treated as exotic curiosities by the public, I asked what the experience is like for her. 'My tics have a physical sensation attached to them,' she says. 'It's often described by doctors as a "premonitory urge", but that doesn't really describe what that feeling is for me. I feel it like itching powder in my blood or the sudden experience of being tugged in a particular direction.'

Vocal tics are influenced by social context. While they can be random, they can also give voice to the unsayable – manifesting as personal insults as well as swear words – in a way that is deeply distressing for anyone implicated in what has been said, but also for the person with Tourette's who may not even share the sentiment coming out of their own mouth. Most tics are not personally insulting, but they can be disruptive in other ways: subversive, surreal, even amusing. 'People think saying "fuck" in the fruit aisle is funny,' Jess says, reflecting on one popular stereotype, 'but that's nothing on the reality of living with Tourette's. I am the one who's being an involuntary Sat Nav while my friends are playing Mario Kart [a racing simulator game] or who's explaining to airport security that there's not really a bomb – or a springer spaniel – in my bag.' Gradually, Jess has come to think of her tics as creative as well as distracting and incorporates them into her stage shows.

There are many intriguing similarities between Tourette's and stuttering, in which I have a personal interest. As well as stuttering, I have experienced motor tics in my face and body all my life – nose twitching, repetitive blinking and sniffing – which I endeavour to conceal throughout the day. While I have never sought diagnosis, the experience of a premonitory urge is deeply familiar to me, and it is not dissimilar to the sensation of an approaching vocal block. I'm struck by the fact that both conditions have elusive origins or causes. Only since the 1990s has there been a growing consensus that they are inherited neurological conditions, suggesting some sort of dysfunction in neural connections. MRI scans show differences in the make-up of the brains of people who stutter or tic: signs of disturbance or 'dodgy wiring', as one neurologist described it to me, which are not there in fluent speakers.

Both are conditions which emerge in childhood, they are around four times more prevalent in males, and in many cases wane after adolescence. And both are dramatically inconsistent and shift according to social context; symptoms come and go for no apparent reason and may disappear altogether. In the same way that a person who stutters can be fluent when performing or speaking a foreign language or on their own, those with Tourette's can often hold tics in reserve then 'release' them later. In one case study, the neurologist and writer Oliver Sacks described a Canadian surgeon who could suspend his tics for long periods of time while performing operations or flying his private airplane.[8]

Like stuttering, tics come and go, affecting people differently day-to-day and over the course of their lives. This dramatic variability in symptoms goes some way to explaining why stuttering and Tourette's have historically struggled to be recognised as neurological disabilities. Such inconsistencies have allowed observers to conclude an element of 'putting it on' as if variability implies voluntariness. Yet it is precisely this variability that renders them so distressing to the individual. Daily life is imbued with an unpredictability that can prove both physically and mentally exhausting, never quite knowing where the boundaries of one's capabilities lie, and always wondering if just a little more effort might enable one to pass as 'normal'.

The coprolalia, echolalia and palilalia of Tourette's syndrome refer to an excess of language. Aphasia, on the other hand, refers to a decline in linguistic ability, affecting speech as well as reading and writing. It is most commonly associated with stroke victims who struggle to speak or whose words are jumbled up. So while stuttering and Tourette's are inherent to an individual, quickly emerging as a child reaches speaking age, aphasia is something that – for the most part – happens to you. It is acquired rather than innate to an individual. It is generally a symptom or consequence of another condition and therefore accompanied by other symptoms: for instance, a stroke victim with aphasia will often have some degree of physical paralysis from the same primary cause. Determining the number of people who have aphasia is therefore a bit of a totting-up exercise, but it has been estimated to affect around a quarter of a million people in the UK.[9]

Like so many names in the world of speech disorders, aphasia is a bit of a catch-all term and there is continuing inconsistency in where the barriers are drawn. Some neurologists believe 'aphasia' should be reserved for linguistic impairments caused by damage to the left hemisphere of the brain; others that it should include damage to the right hemisphere; and others that it should include language impairments caused by dementia and other progressive conditions. When British neurologist Henry Head gathered his case studies of aphasia in soldiers returning from the trenches of the First World War, it was the lack of a general pattern that struck him most of all. 'No two examples of aphasia exactly resemble one another,' he wrote. 'Each represents the response of a particular individual to the abnormal conditions.'[10]

Even the term 'aphasia' covers a wide range of different symptoms, each with their own name.[11] There is *anomia*, which describes a difficulty in recalling the words for everyday objects. There is *paraphasia*, where words which have some logical connection or similarity are confused, like saying 'knife' for 'fork' or 'night' for 'light'. There is *jargonaphasia*, where paraphasia becomes unintelligible: a sentence may make perfect sense to the speaker but comes across as peculiar jargon to the listener. And there is *agrammatism*, where the syntax of speech becomes affected or scrambled, like saying 'I to the house go' rather than 'I go to the house', or is diminished, as in 'house go'.

Aphasia is often accompanied by *apraxia of speech* when a person may know what they want to say but the mouth is unable to perform the task as required. They may say a completely different word or make one up. The tongue and lips may 'grope' to say a word that would once have emerged automatically. While speech-language therapists stress the difference, they are often inseparable. A person with aphasia may both struggle to find the words, then having found them, struggle to get them out. In some cases, people with aphasia also experience stuttering. This sort of stuttering is often called 'neurogenic' or 'acquired' stuttering and, unlike most other forms of stuttering, normally has a clear cause in the underlying stroke, head injury or tumour that brought it on.

Although aphasia generally accompanies other symptoms, including localised paralysis in the arm or leg, people often describe it as far worse than physical immobility because it impedes their ability to communicate. By nature it is something more often described by physicians or family members than by the individuals themselves, something which a few books like *Jumbly Words, and Rights Where Wrongs Should Be: The Experience of Aphasia from the Inside* have tried to rectify. 'Thoughts are clear as bells,' says one individual, 'but come out so muffled and jangled.' 'New words form,' says another, 'lazy words marry each other and the gaps go on.'[12]

The range of severity in aphasia is even greater than with stuttering. 'The individual with mild aphasia may experience only minor problems such as hesitancy in word-finding, or difficulty following a group discussion,' writes speech therapist Gill Edelman. 'In severe cases, unable to understand what is said, to utter more than a few meaningless sounds, or to read or write, the individual may be locked into a private world where normal communication is impossible.'[13] Because it normally occurs later in a person's life it can also be the most debilitating of speech disorders. If you grow up with a stutter, both your personality and relationship with language is indelibly shaped by the experience. Idiosyncratic techniques for managing it are developed over a long time. Aphasia, however, comes out of the blue, ripping away abilities one has always taken for granted.

Fortunately, after the original shock, many with aphasia find the condition improves and they relearn some of their capacity for speech. In *The Word Escapes Me: Voices of Aphasia*,[14] an American woman called Yvonne describes how her aphasia appeared in advance of her stroke. After hopping onto a treadmill at her gym, she tried to make some friendly comment to the person beside her, but her voice came out slurred and garbled. Then she collapsed. As people from across the gym gathered around her, she opened her mouth to speak but no words came. Then she slipped into a coma. 'I was unable to communicate in words,' she recalled, 'I was rendered infant-like again – literally speechless, helpless.' During the long months of rehabilitation, she learned to walk again, although with a limp, and her speech began to return. But recovery was only partial. 'When I try to speak, I'm outed as having trouble talking. "Ss" become "Shs", "single" becomes "shingle", "spell" becomes "smell", and so forth.'

For many with aphasia, this rehabilitation is incredibly slow and arduous. Few ever recover the capacity for speech they previously enjoyed. Since language is central to how we think and express ourselves, this transformation can be experienced as a fundamental change in personality. This is often, but not always, described in negative terms. But while many talk about the overwhelming frustration and even anger, particularly in the early stages, others (admittedly, a minority) describe a necessary reassessing of priorities and values, a sense of enhanced empathy and eventual peace of mind.

The fourth, and last, of the most common speech disorders is dysarthria. However, it is not a word people use much. Even more than aphasia, it is often one symptom among many of an underlying condition. The term describes the distorted articulation, the slurred or slow speech, of those with cerebral palsy, muscular dystrophy, Parkinson's disease and any other condition which causes damage to the nerves or muscles involved in speaking. It can also include stuttering-like symptoms. Understanding someone with dysarthria can be difficult and, as a result, intelligibility is one of the key indicators in determining an assessment. Because of the wide range of causes, we do not know exactly how many people have dysarthria, although if one considers that Parkinson's is thought to affect 127,000 people in the UK,[15] while around 166,000 have cerebral palsy,[16] then its frequency may be as common as aphasia.

While few people see themselves as having dysarthria as such, its presence within a broader condition can often prove the most painful symptom. Jamie Beddard is a playwright, actor and director who was born with cerebral palsy. He can walk, but increasingly relies on his wheelchair now he is in his fifties. His speech, in his own words, is both 'guttural' and 'hard to understand', although it is easy enough to follow with a little concentration. We first became acquainted through our work in the arts, often finding ourselves at the same events. After a year or two of knowing one another by sight, I realised I was doing what so many do: avoiding direct conversation because I was afraid of not understanding and causing offence. So I invited him for coffee.

'I used to say the only thing that bothers me is my communication,' he tells me. 'My speech has been the biggest element of my disability.' To enable better communication, Jamie puts a lot of physical effort into articulating. As a result, the act of speech can be physically exhausting, requiring a certain economy over what he does and doesn't say. While dysarthria often lacks the dramatic variability of Tourette's or stuttering, it still changes from day to day. 'It's not consistent,' he says. 'If I'm relaxed, it's probably a little better. When I'm nervous, it's a little bit worse. When I've had two beers, it's a little bit better. When I've had four beers, it's a little bit worse. Basically, the more relaxed I am the better.'

While people with cerebral palsy are normally born with dysarthria, there are many others who acquire it later in life. Motor neurone disease (MND) refers to neurodegenerative disorders affecting the nerves in your brain and spinal cord, including the capacity for speech. It is rare, affecting around 5,000 people in the UK.[17] Fortunately, there is greater awareness than there used to be because of the distinctive image and voice of scientist Stephen Hawking who lived with it for over fifty years. But Hawking was the exception: life expectancy following diagnosis of MND averages between two and five years.

My cousin Gilly Truman was diagnosed in her early thirties. Seven years on, she now uses a wheelchair and her ability to communicate is in what she calls the 'transition' phase with an increasing dependence on her augmented and alternative communication (AAC) device. 'I have to fill in a form every quarter in my clinic about dysarthria,' Gilly tells me. 'The questions ask about how difficult I find talking. I say don't ask me about how difficult it is. It's more the effort. If I've had a massive weekend of socialising, I find I just can't talk any more. I can't make myself heard in restaurants anymore so I find it easier to socialise at home.'

Like Beddard, Gilly emphasises the variability and inconsistencies of life with dysarthria. 'If I don't know someone, I find it harder to speak. The effort is tenfold. Or when I'm emotional, like when I'm talking about MND, my mouth gets a bit breathy. Basically, I can't hide any emotion or anger in my voice.' But the greatest inconsistency stems from the unavoidable process of transitioning, week on week, year on year. 'It will only get worse,' Gilly says. Increasingly, she is using EyeGaze, a sophisticated technology that tracks the movement of her eyes on a computer screen to spell out and pronounce words. This is partly because it can be a welcome break from the effort of speech, but also because she knows EyeGaze, rather than her worsening dysarthria, will be the voice of her inner thoughts and feelings for the rest of her life.

Together, stuttering, vocal tics, aphasia and dysarthria encompass most cases of pathologically disordered speech. But there are also conditions which affect speech in other ways. The most common of which are voice disorders, or dysphonia, caused by abnormalities in the voice mechanism in the larynx. The causes are manifold: from certain types of cancer and multiple sclerosis through to exhaustion, smoking and the temporary symptoms of the common cold. People with such disorders may be perceived, by themselves as well as others, as having voices which are too low, high, quiet, loud, monotonous, rough or hoarse. These symptoms may be accompanied by a physical difficulty in speaking.

One high profile figure with dysphonia is broadcaster Nick Robinson who describes his voice troubles following an operation for lung cancer in his book *Election Notebook* (2015). Because there is a strong element of subjectivity in diagnosing a voice disorder (one person's too hoarse, is another's just right), studies on the prevalence of dysphonia in the UK vary widely, although one estimate suggests around 2.5 per cent of the population.[18] As with stuttering, I think it is an individual's own perception of struggle that should determine a diagnosis most of all.

There are other conditions, but they are far rarer. Spasmodic dysphonia, for instance, is a neurological disorder in which the muscles that generate a person's voice go into periods of spasm. This creates breaks or interruptions in the subject's voice as frequently as every other word, making speech very difficult to understand. Dysprosody, known as foreign accent syndrome, is an extremely rare neurological speech disorder, usually caused by brain damage from a stroke or tumour, in which the variations in pitch and timing control go askew: people know what they want to say but can't control the way the words come out of their mouths. Often this results in the subject speaking in a pseudo-foreign accent.

There are also conditions which fundamentally change the way an individual communicates but are not speech or voice disorders as such. Selective mutism is a severe anxiety disorder that affects around 1 in 140 children.[19] They find themselves unable to speak in certain social situations and, left untreated, it can continue into adulthood. 'Having selective mutism can feel like you're living your life in a box,' writes Sabrina Branwood, a thirty-something woman from Rochdale, in an interview with the BBC. 'The box is see-through so you can see out and hear people, but you can't leave no matter how hard you try. You can shout inside the box as loud as you like but nobody can hear you. They can't hear you cry when you're hurt or scared.'[20] One famous person with Selective mutism is enviromentalist Greta Thunberg who has described her anger at the climate crisis as an overwhelming force that makes her speak in spite of her condition.[21]

Finally, there are those deaf people who use speech but struggle to articulate their words in a way many can easily understand. Increasingly, such 'oralism', the term used to describe the practice of teaching deaf people to communicate using speech and lip-reading, is considered old-fashioned. It is rejected as something imposed by intolerant societies rather than chosen and the use of sign languages is on the rise. In the 2011 Census, for instance, 22,000 people in England and Wales reported using Sign as their main language.[22] However, the term 'speech disorder' is inappropriate for the communication methods of the deaf. The difference comes down to choice. While those with speech disorders experience a consistent and obtrusive struggle in speaking the words they want to say, deaf people may use verbal speech or sign language, or a combination of both, in a way that is 'fluent' and effortless.

So far, I have described some of the main speech disorders on their own terms, as unique conditions with occasional overlaps. Stuttering is an impairment in fluency; aphasia in use of language; dysarthria in articulation; and vocal tics in intent, with unwanted words interrupting speech. For good reason, these conditions are generally, and appropriately, considered in isolation, and the literature around them is specific to each. Yet there is another way of looking at it. Rather than stressing the differences between speech disorders, we can emphasise what they have in common and think of them as a family: a sprawling one with different personalities, but a family nonetheless.

This is not an arbitrary exercise, but one that is critical for any serious attempt to tackle widespread discrimination against those who have them. Considered individually, they can appear rarified or obscure conditions. As a result, the case for research funding has sometimes been hard to make, and campaigns to tackle social prejudice have rarely reached critical mass. But if we consider them as a family, we are forced to acknowledge that they impact all of our lives. We may not be one of the millions who have them, but somebody in our life almost certainly does. What is more, we may develop one in years to come; whether the aphasia of a stroke or the dysarthria of a late-life condition like Parkinson's disease. It is in our interest, and that of people we care about, to ensure that such conditions are better understood and cease to be discriminated against.

Perhaps the most significant breakthrough for this viewpoint is the increasing scientific consensus that most speech disorders are neurological in nature, resulting from disturbances to the brain and nervous system. This is an idea we are still getting used to. Up until the end of the twentieth century, both stuttering and vocal tics were widely considered psychological complaints, putting them in an entirely different medical category to aphasia and dysarthria. Going further back, stuttering was explained in physiological terms, as the result of a large tongue or lack of moisture in the body.

Brain scanning, however, suggests that while the causes of a speech disorder can range from cerebral palsy to stroke as well as the 'enigma' of stuttering, in each case there is some difference in the circuitry of the brain. This would explain why many people experience more than one speech disorder. For instance, neurogenic, or acquired, stuttering often co-occurs with aphasia and dysarthria.[23] It would explain too why variations in dopamine, a chemical neurotransmitter in the human brain that sends signals between nerve cells, impact so many speech disorders. Drugs that suppress dopamine have proved effective in treating Tourette's syndrome and stuttering, while those which enhance it have diminished the symptoms of dysarthria in conditions like Parkinson's disease.

The gradual streamlining of speech disorders into a neurological framework has been accompanied by a similar streamlining of treatment. In the past, depending on your condition, you might be treated by a surgeon, a psychoanalyst, or just written off as a hopeless case. Today, if you seek a diagnosis or treatment for symptoms of aphasia, dysarthria, cluttering or stuttering, you will almost certainly find yourself in the company of a speech-language therapist. This is entirely appropriate because the techniques a therapist uses often apply equally well to different disorders. Yet such mainstreaming of the profession was hard won.

Speech therapy developed as an outsider discipline, frequently dismissed by the medical establishment right into the twentieth century, but gradually gained currency because of its ability to mitigate the symptoms and negative experience of speech disorders more effectively than other practices. The techniques used can seem deceptively simple, whether teaching better articulation or breathing, but it often overlaps with psychotherapy, helping a demoralised or despairing client to self-confidence and better understanding. Today, the experience of speech therapy – often, but not always, a positive one – is something that people with many different types of speech disorder share.

But more than their shared neurological origins and similar methods of treatment, I believe what unites all speech disorders is their relationship as a minority 'other' to fluency or 'normal speech'. Anyone with a speech disorder experiences some form of disruption between what they want to say and their ability to say it. The person with aphasia or a stutter, or the coprolalia of Tourette's finds that words either won't come out or do so differently, or are interrupted by other words that seem to appear from nowhere. Such people have a fundamentally different relationship with language than fluent speakers do; it is something to be distrusted as much as enjoyed.

This is why I find Ngũgĩ wa Thiong'o's description of a 'malady of words' in his novel *Wizard of the Crow* so powerful. When I was a boy, and long into adulthood, I believed I was cursed by a rare affliction. I had little idea how many people pathologically struggle with their speech. For me, Ngũgĩ's vision of a nation where millions are afflicted by a 'malady of words' is more than a metaphor for political voicelessness, but a description of reality. The phrase cuts through the medicalisation, the terminology, the cordoning off of myriad conditions, and reminds us that collectively a significant proportion of our population struggles with speech and feels humiliated by it.

'Humiliation can leave scars for life,' Ngũgĩ told me on that winter's afternoon in Venice:

> You can cover it, but the humiliation can be internalised. You begin reacting against certain things, not because you are consciously thinking about rejecting this or that. It becomes almost like the way we avoid spaces of pain because we like to inhabit places of comfort. So if you've been humiliated in relation to language, even if just the register of your voice, without realising it you don't want to have anything to do with that register.

As he said this, I realised that the physical symptoms of speech disorders – the stuttered sounds, unexpected interjections, lost or mumbled words – are not the defining qualities of those conditions. Running beneath them, and far more overwhelming for the individual, are the negative emotions: the feelings of isolation and shame that shape a person's life. On the flight home from our meeting, I opened the copy of his prison memoir, *Wrestling with the Devil*, which I'd hurriedly asked him to sign as I left. Above his name, he'd written: 'In solidarity in the struggle for the right to one's language.' Not a commiseration then, but a call to arms.

2

The Mouth Trap

In the winter of 1878–79, the novelist Henry James famously dined out 140 times in and around London. As a young man off the boat from America, he was fascinated by the glamour and sophistication of the Old World. His notebooks and letters are full of observations about people's speech and conversation. He reflects on the 'high superiority of French talk', encounters 'one of the most charming and ingenious talkers I ever met', is relieved at a dull soirée to stumble upon 'a flowering oasis in conversation sands', disapproves of an acquaintance who is 'rendered more inarticulate than ever', and worries that his own standards of 'what makes an easy and natural style of intercourse' might be dropping.[1] This fascination with the art of human conversation fills the novels and stories he is best known for, including *A Portrait of a Lady*, *Daisy Miller* and *The Wings of the Dove*.

His linguistic snobbery could also make him a bore. Having spent much of his life avoiding lower-class establishments, he unwittingly found himself in an Upper East side café in New York (the city had changed a great deal since his previous visit) where he felt himself in 'the torture-rooms of the living idiom'.[2] The experience was so traumatic, he addressed it head on in a lecture on 'The Question of Our Speech'. Speech is sacred, he said, and warned that 'the human side of vocal sound' was being corrupted by slovenly speech and kept 'as little distinct as possible from the grunting, the squealing, the barking or the roaring of animals'.[3] When the young Winston Churchill met James he couldn't resist winding him up by using as much slang as possible.[4]

Such preoccupation with language, and its correct usage, is understandable in a novelist determined to write both exquisite prose and capture the speech patterns of the sophisticated Londoners and Americans he encountered. But something else motivated it too, a secret he shared with very few.

That there was something strange about James's own speech was impossible to ignore. And just as he wrote about the speech of others, many people wrote about his. James's manner of speaking was long-winded, full of pauses and circumlocutions, and it proved divisive. Some found it irritating: 'I don't think he talks remarkably well,' wrote William Hoppin, an American diplomat in London. Constance Fenimore Woolson, a fellow novelist, tactfully struggled to explain his 'unusual flow of language', while the poet Ezra Pound remembered him 'weaving an endless sentence'.[5] The young Virginia Woolf left a satirical sketch of their encounter:

'My dear Virginia, they tell me – they tell me – they tell me – that you – as indeed being your father's daughter nay your grandfather's grandchild – the descendant I may say of a century – of a century – of quill pens and ink – ink – ink pots, yes, yes, yes, they tell me – ahm m m that you, that you, that you write in short.'[6]

Like most men of his time, James was not good at talking about his deepest feelings. He was almost certainly gay, but seems to have chosen the (arguably) less complicated life of celibacy. In the same spirit, he never acknowledged his stutter. But stutter he did. Some of those who encountered him hint at it. His nephew noted his uncle's 'perpetual vocal search for words even when he wasn't saying anything'. Urbain Mengin, a French poet, noted that 'he speaks with a slight hesitation [a word often used in polite circles at the time as a euphemism for stuttering], repeating the first syllable of certain words.' Miss Weld, James's typist, also acknowledged the 'hesitation', but promptly adds that it 'was simply nervousness and vanished once he knew you well'.[7]

One of the few people James confided in was fellow writer Edith Wharton, who had a knack for drawing people out:

His slow way of speech, sometimes mistaken for affectation – or, more quaintly, for an artless form of Anglomania! – was really the partial victory over a stammer which in his boyhood had been thought incurable. The elaborate politeness and the involved phraseology that made off-hand intercourse with him so difficult to casual acquaintances probably sprang from the same defect. To have too much time in which to weigh each word before uttering it could not but lead, in the case of the alertest and most sensitive of minds, to self-consciousness and self-criticism; and this fact explains the hesitating manner that often passed for a mannerism.'[8]

What is most striking about Wharton's claim is the extent to which it suggests Henry James's life was shaped by a speech impediment few ever actually heard. It determined not only his elaborate and long-winded way of speaking, but the way he thought about himself. Above and beyond Wharton's theory, I think it may have informed his obsession with the way language is used by others; and, as a writer who dictated many of his novels, it is present in many of the books he is most famous for (I will say more about this later). But in all this there is nothing unusual.

We define and diagnose speech impediments by their overt symptoms: the strangulated repetitions and blocks of stuttering, the outbursts of vocal tics, the distorted articulation of dysarthria, and the pauses and mala-propisms of aphasia. Yet for anyone who has a speech disorder, it is not the sound of their condition that most impacts their lives, but everything that follows. Speech-language therapists sometimes use an iceberg metaphor to describe this.[9] The act or sound of a speech disorder is just the part others are aware of, if at all, but underneath it are complex behavioural and psychological responses. What is experienced by others as a disorder of speech or even just an unusual manner of speaking can shape an individual's entire personality.

As an overt, and then interiorised, person who stutters,

I developed techniques for managing my speech from a very early age; not only avoiding problematic sounds and words, but situations where I might be required to use them. The experience of being and feeling humiliated damaged my self-esteem, ultimately impacting the choices I made in life. I'm one of the many who, like Henry James, manage to pass as fluent, if sometimes unusual, speakers. We keep our disorder invisible but at great cost, restricting and narrowing our lives to do so. In this regard, I both do and don't have a speech disorder, for stuttering is the thing I have spent a lot of my life energetically not doing.

Of course, there are many others for whom concealment is not possible. The sense of shame they feel and are made to feel by others can result in low self-esteem, isolation, loneliness and, in extreme cases, dissociative disorders and suicidal feelings. One of the reasons why I believe we should talk about speech disorders collectively is because their impact on an individual is primarily psychological. There is a consistency of negative feelings and avoidance behaviour that applies to those across the spectrum, whether they stutter, tic, struggle to articulate or to recall words. By understanding the way these disorders impact on an individual's life, I hope we can respond and better support those affected.

For people with a speech disorder that manifest itself in early childhood, like stuttering, vocal tics or certain forms of dysarthria, it can be the reactions of others that inform their sense of something being wrong rather than any innate consciousness of something awry in their speech. 'Adverse listener reactions can play a part,' writes linguist David Crystal. 'A typical example is when parents prematurely correct their children for non-fluency, or become impatient when their child is non-fluent; this causes insecurity and anxiety, which in turn causes further growth in the non-fluency.'[10]

Shame, it seems, often starts at home not out of malice, but from love. Parents want their children to lead 'normal' lives after all. 'I know people whose families are uncomfortable with them ticcing,' artist Jess Thom tells me, 'and there's an unconscious pressure to suppress which can have a really negative knock-on effect on their mental well-being and their mental health.' If symptoms continue, parents may seek a diagnosis. This can result in a child being told they have a problem before they are even aware of having one.

Francesca Martinez is an actor and comedian with cerebral palsy. When she was a baby, doctors told her parents she was physically and mentally disabled and would never lead a normal life. 'I'm not quite sure what a normal life is,' she writes in her memoir *What the **** is Normal?!*. 'What I am sure of is my bemusement now at the ease with which these professionals make such weighty pronouncements. Words that take seconds to utter and decades to cast off. Unknown to me then, they disappeared into the ground around me and, over time, would emerge as the bars of a cage, hemming me in from the outside world.'[11] As she grew older, her parents encouraged her, lovingly but forcefully, to overcome her slurred speech, although it wasn't a problem she recognised. 'I couldn't see why I needed to do exercises at all,' she writes. 'What was the logic in practising certain words to improve my speech when I talked perfectly already? A fact clearly demonstrated by the clear voice I heard every time I spoke. Or in the suggestion that certain mouth exercises might help me not to dribble.'

While self-consciousness about speech and labelling can aggravate a problem, the pioneering American speech therapist Wendell Johnson went one step further. Stuttering, he claimed, begins not in the child's mouth, but in the parent's ear. The act itself is nothing more than 'the simple repetitiousness of preschool-age children.'[12] It becomes a clinical problem 'not before being diagnosed, but after being diagnosed … The more anxious the parents become, the more they hound the child to "go slowly", to "stop and start over"… the more fearful and disheartened the child becomes and the more hesitantly, frantically and laboriously he speaks.'[13]

To some extent Johnson was right: 5 per cent of children stutter at some point, generally between the ages of three and six. And there is surely some significance in the fact that children who stutter have faster speech rates on average than children who do not. For many, it may be a verbal version of the stumbling they do when learning to walk and run, yet few if any parents fixate on a toddler's tripping over as a sign of congenital lameness. In fact, 80 per cent of children who stutter become fluent speakers and it is sometimes hard to tell whether speech therapy alleviated or exacerbated the problem.[14]

In the 1930s, Wendell Johnson put his theory to the test in the now infamous 'Monster Study'. Under his supervision, graduate student Mary Tudor selected six children between the ages of five and fifteen at an orphanage in Iowa for an experiment. The orphanage was important because she wanted children unencumbered by protective parents. 'You have a great deal of trouble with your speech,' Tudor told each child, all of whom were already vulnerable from a life in care. 'These interruptions indicate stuttering. Don't ever speak unless you can do it right. Whatever you do, speak fluently and avoid any interruptions whatsoever in your speech.'[15]

Tudor studied their behaviour in the following months and found that they spoke less, more slowly and with greater hesitations. Their behaviour changed too; they became shy and easily embarrassed children. Years later, Franklin Silverman, a revered speech therapist and one of Johnson's students, recalled that Tudor continued to visit the children in following years out of a sense of responsibility because some of them had actually developed stutters as a result of the 'experiment'. 'The implications of the findings seem clear,' Silverman wrote. 'Asking a child to monitor his speech fluency and attempt to be more fluent can lead to increased disfluency and possibly stuttering.'[16]

Knowing all this now, it is hard not to look back and wonder to what extent my own stutter was shaped by my family and nursery teachers. My earliest memories aren't of actually stuttering, but my family's reaction to it. Since stuttering runs in families (and there is even some evidence of a stuttering gene), parents who experienced it themselves may be particularly sensitive to any telltale signs in a child's speech. They may hear an echo of their own affliction in the perfectly normal repetitions of speech development and seek a diagnosis and treatment before it is necessary.

My mother stuttered from school age well into her

twenties, an experience she remains sensitive about to this day, so it is understandable she was highly attuned to any obstacles in my own speech. But I am only partially convinced by Wendell Johnson's theory. He was working at a time when the influence of psychoanalysis was at its peak, and there was a tendency to over-emphasise the role of neurosis in many pathological conditions. While I believe it possible in some extreme cases that a child can start to stutter purely through suggestion, the evidence of brain scanning suggests many more are born with a disposition that may be lessened or intensified, but not caused, by how it is treated by the adults around them.

Johnson wrote about stuttering, but a hereditary pattern has also been found in Tourette's syndrome, suggesting family influence could play a similar role. Johnson's insights are clearly less relevant to the speech disorders arising from conditions like cerebral palsy, where symptoms may be evident long before the speech development process begins. Even then, as Francesca Martinez describes, the process of diagnosing a condition is a sensitive one that can affect a child's confidence for the rest of their lives. In all cases, I think parents would do well to worry less about disordered speech in a child and consider it instead as a perfectly normal process in speech development, even if 'normal' is never going to be the same as for other children.

The loving concern and implicit pressures of anxious parents may generate a sense of shame, but it is only aggravated when that child is thrown into a larger and less sympathetic social environment. Joshua St Pierre, a Canadian philosopher and speech activist, vividly remembers the frequent humiliations of school. 'There'd always be shame and embarrassment,' he tells me on the phone. 'And after these awkward social situations, I'd mutter things like "stupid Josh, stupid Josh, stupid Josh" as a way of mitigating that awkwardness.' The Irish novelist Colm Tóibín tells me about 'a little fucker called Titch Hogan. He would follow me home going "duh-duh-duh-duh" the whole way. I put his mother into one of my books.' The American writer Darcey Steinke writes about the occasion some kids from her school tossed a book called *The Mystery of the Stuttering Parrot* onto her family lawn.[17]

'My absolute worst nightmare,' the poet Owen Sheers tells me, 'was the idea of reading round the class in English because you can see your turn coming: you can see the paragraph with the words – and blocks – you're going to have.' Oliver Dimsdale, long before he even dreamed of being an actor, remembers 'looking around at John or Fred who was next to me, nonchalantly reading line after line, and thinking you lucky, lucky bastards. How on earth are you able just to pick something up without thinking about it and read fluently? It would get round to me and I'd spit out two or three sentences. More often than not, I'd run out and find a place to have a cry.' My mother experienced this situation differently: for her it was about exclusion rather than humiliation. Even today she recalls bitterly the way her teacher in the 1950s wouldn't let her speak, simply saying, 'I think we'll skip Anne and pass on to the next.'

It was partly to counter the misconceptions of teachers and the taunts of children that a group of worried parents formed the Tourette Association of America in the early

1970s, supporting new medical research as well as awareness-raising campaigns. Fifty years on, the TAA continues to gather the experiences of parents and children impacted by a condition that remains widely misunderstood. One mother describes her horror at hearing her seven-year-old announce that he wanted to kill himself. 'Here was my son telling me in the clearest words possible that he felt he had no value,' she writes, 'that having Tourette's syndrome meant his future would never be equal to his peers. He was giving up ... I thought of the mistakes I'd made before his diagnosis, remembered every time I'd begged, "please, just stop!" I thought about the times he'd been removed from class, punished, and sent home.' One man recalls his worst anxieties about Tourette's being around the age of eleven, 'when establishing your "cool" took a lot of hard work and one little snag – like tripping down the stairs, spilling soda on your pants or, God forbid, an episode of uncontrollable sounds and movements – could set you back years, if not eternally.'[18]

It is all too easy to blame children and teenagers for humiliating schoolmates who struggle with speech, but I believe our education system promotes such behaviour. Schools are built around relentless displays of public speaking: reading aloud in class, assembly presentations, plays, debates, oral examinations. In all of this, they do little to accommodate those, and there are many, who do not respond well to such tasks. Not only people with speech disorders, but introverted or easily embarrassed children.

Despite being diagnosed very young as a person who stutters, I was never exempted from such rituals. At the age of nine, I occasionally pretended to forget my own name in class roll-calls to avoid stuttering on the troublesome 'J'. Ten years later, I failed the oral in a French exam because I blocked so severely throughout the test. In between those two events is a long line of humiliations in which I recall my teachers in part as threats who could expose me at any moment through the tasks they set.

The argument that such tasks are a necessary preparation for life to come is half true at best. Few people embark on careers that require anything close to the amount of public speaking they were compelled to do at school, and there are many whose experiences were so bad that they choose careers precisely to avoid it. An education system developed with an awareness of the number of children who experience speech disorders would place less emphasis on performative speaking, ensure what tasks they do participate in are not humiliating for them, and place equal attention on other forms of communication that may play better to their strengths.

Although many speech disorders are developmental and dissipate over time – only 20 per cent of children who stutter or have vocal tics continue to do so in adulthood – the damage may be already done, with self-respect and social confidence forever diminished. For those who continue to struggle, a pattern of avoidance and low self-confidence is already set. Life choices, whether careers or relationships, may be determined by the desire to blend in.

Jim Smith, a leading UK scientist who happens to stutter, tells me that he only went into his field because of a misguided idea of the 'lone boffin'. He had no idea of the amount of lecturing, seminars, committee meetings and policy speeches he would have to do. In her early twenties, my mother was a junior reporter with the *Sydney Telegraph*. She would never ask a question at a press briefing because you had to say your name first, something which caused her to stutter. In the aggressive culture of journalism, she was reluctant to expose herself to public humiliation. She even adjusted her first name from 'Anne' to 'Annie' because the 'e' sound gave her a bounce into 'Woodham'. After a while, the hopelessness of being a news reporter who wouldn't open her mouth at junkets became too much and she went into research-oriented features writing instead. It still involved asking questions, but without a dozen other journalists listening in.

As a young man, I made similar choices. I considered, and dismissed, professions that revolved around performative speaking. The law, politics and journalism were intriguing but doors better left closed. Instead, I went into documentaries, an art form which generally tells stories using the voices of others. It turned out I had a limited idea of how they are made. My early jobs as a researcher involved endless cold-calling while sitting in an open plan office, which is a trigger for stuttering. I ran up vast bills with my late 90s brick-shaped mobile phone making calls in the stairwell or outside, for which I could never claim reimbursement as I had no plausible reason for not using the perfectly functional phone on my desk. Another accessory was a pack of beta blockers which I drew upon before important pitching meetings or interviews to help me relax and therefore speak a little more fluently.

Although there is certainly an element of self-censorship

in such career decisions, discrimination remains a real problem. Francesca Martinez writes that, as a comedian, she has been dropped from BBC radio shows because of her 'funny voice'. Walter Scott, a civil servant in the Ministry of Defence, tells me he was rejected for a scholarship with the Armed Forces as a young man because of his disfluent speech. 'I had subsequent experiences in job applications in my early twenties,' he says, 'where I was rejected because I stammered. Eventually, I entered the civil service by applying for a job nobody else had applied for. I have chosen my career path carefully, generally avoiding jobs where I thought there'd be other people applying. Why would you take the candidate who has a stammer?' And there are many with acute aphasia or dysarthria, or who communicate with Sign, who cannot get any work at all.

Discrimination in the workplace may not be as overt as that of the schoolyard, but it is no less damaging to the esteem and potential of an individual. Think of how often job adverts specify the need for 'fluent and effective' communications. And it is still widely assumed that there are certain jobs that are entirely incompatible with speech disorders: jobs which require giving quick orders (traffic control), or representing a company to the outside world (public relations), or talking to customers (sales), or cold-calling (research). Each employer will have their own set of positions for which a basic mastery of fluency seems essential. Case-by-case, there are always arguments and justifications which seem sound, but looked at collectively people with speech disorders unwittingly form a social caste better suited – in the eyes of society – to manual or lower-income desk jobs. The problem isn't so much one of wilful prejudice, but of ignorance. There are, in fact, very few jobs that people with speech disorders cannot do, particularly since technology has provided devices and apps that can compensate for the words an individual struggles with. Social attitudes are not going to change overnight but a powerful first step would be for companies to build an awareness of speech disorders, as they do with other types of diversity, into their employment policies to avoid discrimination. This is currently patchy at best and it is usually down to an individual whether they consider and are willing to describe their disorder as a form of disability; something which in theory protects them under the Disability Discrimination Act.

While most speech disorders develop in childhood, there are many which emerge later in life. In such cases, feelings of shame are no less pronounced. The person with aphasia following a stroke might feel half the person they were before, reduced by the loss of speech to a state of near-infancy. 'Just because I can't speak like them and I look a bit different,' writes a man called Jim who experienced a stroke at twenty-nine, 'they turn away, ignore me or speak to my wife or speak to somebody else.'[19]

Those who find their ability to articulate affected by Parkinson's disease or motor neurone disease are acutely aware of slipping, in the eyes and ears of those they encounter, from the status of the able-bodied to that of an invalid. My cousin Gilly tells me of the shock of becoming socially invisible, both because of her wheelchair and her dysarthria. 'When your voice goes, people assume your brain has gone too,' she says. 'A few months ago, I went into a clothes shop with a girlfriend. The shop assistant would only talk to her. Eventually, my friend looked at me and said, "What clothes do you want?", and I could see the shop assistant realise that I did have a brain.'

Faced with the prejudices of an intolerant society, and the sense of shame which that often creates, people with speech disorders develop a complex range of tactics for managing their speech. Those who can will try to conceal a condition for as long as possible. This is particularly true of stuttering which, unlike vocal tics, aphasia and dysarthria, is often self-contained rather than a symptom of a more multifaceted and widespread condition like Tourette's, stroke or cerebral palsy. As a person grows to understand the words or situations that affect them most, they also become more adept at avoiding them.

Like most people who stutter, I developed these techniques through my teenage years. I did so instinctively, grasping at anything that might get me through or away from a difficult word. Only as an adult, when I finally began talking to others and reading books on the subject, did I realise how universal these 'dodges' are, effectively amounting to a grammar of stuttering. There is *word substitution*, where you simply replace a difficult word with a synonym (and this might explain in part why there are so many words for stuttering, including the softer, less plosive 'stammering'). There is *deliberate hesitation*, where you wait until a word can be more effectively broached, much in the way George VI did in his radio broadcasts; and *pitch modulation*, where you adjust the speed and tone of your voice, generally by slowing and flattening it, so it is less bumpy. There is also *deliberate repetition*, where a speaker bounces into a difficult word, but without the involuntary grimaces of stuttering.

Used together, these techniques can result in a highly idiosyncratic way of speaking, like the descriptions of Henry James's speech. Looked at again, with an awareness of the full range of techniques used, and considering James's intention to avoid actually stuttering, his speech no longer seems strange but extremely logical and structured. The American journalist Elizabeth Jordan befriended James late in his life and wrote an article, published long after his death, describing the way he spoke. In it she recalls that he broke his sentences into 'little groups of two, three, or four words', just as speech therapist Lionel Logue taught George VI to speak in three-word breaks. In a transcript of one such sentence, she also captures all the blocks, elongated vowel sounds and varying rhythms we have identified as common tactics:

> Eliminating – ah – (very slow) eliminating – ah – eliminating nine-tenths – (faster) nine-tenths of-of-of (very fast) what he claims (slower) of what he claims – of what he claims (very slow) there is still – there is still – there is still (very much faster) enough – left – e-nough left (slower) to make – to – make – to – make – a remarkable record (slow) a remark-able record, (slower) a remarkable record (very slow).[20]

When I attended Carolyn Cheasman's course at City Lit for 'interiorised stammering' in my early thirties, I didn't even know such a phenomenon existed beyond my own secretive and shameful behaviour. I was astonished to find that nearly all the people in my group, about ten in all, were only 'out' to family and close friends. There was one man who had even managed to keep it a secret from his fiancée. 'It goes far beyond physical acts,' Patrick Campbell, an alumnus of City Lit and co-author of *Stammering Pride and Prejudice,* tells me. 'People who stammer talk about not being able to say what they want to say; changing their words; not speaking out when they want to speak; being the quiet one in the room; shaping their whole life around this thing. That is what stammering is, much more than the physical issue it causes in your mouth.'

Patrick's comments remind me that although I use words like 'tactics' and 'strategies' to describe the ways in which people who stutter conceal or mitigate their disorder, it implies more agency than they feel. Interiorised behaviour ceases to be a conscious choice but second nature. The words I speak today are vetted by an internal and unconscious mechanism I rarely pause to acknowledge. In the same way, I do not think Henry James's strange way of speaking arose from him consciously avoiding particular words, but from a long and deeply ingrained habit of which he was no longer aware.

Avoidance strategies are by no means restricted to stuttering. While many with aphasia, dysarthria or vocal tics have speech that is palpably disordered, affecting every word they utter, many find their condition is both moderate and variable: with great effort and numerous tricks they can also pass as something close to 'normal'. In my experience, the more an individual is able to appear fluent, the more they may attempt to do so. Ben Brown, an American with Tourette's syndrome and the founder of the Tourette's Podcast, writes how he 'kept a lid on my TS and crafted ways to mask my tics or blend them into normal routines, even as the subterfuge could be exhausting and often unsuccessful'.[21]

Similarly, those with degenerative conditions that affect speech will often try to conceal their condition for as long as possible. Writer and documentary filmmaker Jon Palfreman describes the patient support group he attended after being diagnosed with Parkinson's disease. There was a long conversation about 'coming out'. One-by-one, patients admitted they hadn't told anyone but their closest family about their condition. A woman in the group warned them that 'if you kept matters secret, people might interpret your behaviour – the slow movements and slurred speech – as something else. "They might think you'd been drinking … that you're an alcoholic."'[22]

The actor Michael J. Fox has talked about his desperate attempts to drug his Parkinson's into submission long after diagnosis. 'I can vividly remember all those nights when the studio audience, unknowingly, had to wait for my symptoms to subside,' he wrote, recalling his third season in the sitcom *Spin City*. 'I'd be backstage, lying on my dressing room rug, twisting and rolling around, trying to cajole my neuroreceptors into accepting and processing the L-DOPA I had so graciously received.'[23]

Attempts to conceal a speech disorder, or the condition causing it, often largely succeed in their aim, but they come at a high cost. 'You can imagine the pressures of having to keep those levels of secrecy going," says Willie Botterill, one of the founders of the Michael Palin Centre for Stammering Children, and my speech therapist for many years:

> I had a woman who worked for the health service who had just got to the point where she was having to make speeches to quite big audiences about technical stuff. It is very difficult to find your way around a specific word which applies to something there is no other word for. She came to me because she was having terrible migraines as a result of the stress of trying to keep it all secret.

Betony Kelly is a civil servant who decided in her mid-thirties to stop hiding her stammer. She wrote a blog on the civil service website announcing her decision and why. 'Depending on whether it was a good or bad day,' she tells me when we meet in a cafe in Whitehall, 'between 10 and 30 per cent of my brain was taken up managing my fluency. When you start to speak and you have a stammer, the seamless connection between your thoughts and your mouth just isn't there.' As a result, a disfluent speaker is never quite in the moment. However present they appear, an inner process is at work: scanning ahead; determining when to engage and how; sifting words like sand to remove the clunky grit.

'In many ways you have to be able to see other paths, other ways of expressing yourself,' Kelly says. 'All those strange things we do, like pretend we've forgotten a word and get someone to spell it for us. Or speak in a different accent. All those little playful things just to get through a sentence.' When Kelly came out at work, she didn't feel ashamed but liberated from the 'huge emotional and physical drain of trying to produce fluent speech'. She accepted what so many deny: that fluency-performing techniques have the adverse effect of worsening rather than improving communication. The person who endlessly substitutes words to avoid stuttering confuses a listener far more than one who allows those same passing blockages to occur.

Of course, there are many whose speech disorder cannot be concealed. For them, a different set of tactics needs to be cultivated. These are about mitigation; rendering themselves and their speech more palatable to a fluent majority. Actor and director Jamie Beddard describes how his cerebral palsy requires him to disarm a listener just to enable a simple conversation. He has a sophisticated strategy and set of techniques for achieving this. 'If I can make people relax, they are able to understand me. But people who think, "Oh my God, I don't know what he's saying" are unlikely to focus to do so.' Humour is an essential tool. 'Often I say, "if you don't understand me – tough!" and they understand I am taking the piss.' Laying out the ground rules is important too. 'I tell them it's okay to say "what?" I don't mind repeating myself again and again. The moment I give permission to not understand, it makes it a lot easier. I prefer that to when people pretend to understand me when they clearly haven't.'

The imperative to conceal, to moderate, to explain and apologise joins with the shame an individual feels and exacerbates what is undoubtedly the dominant psychological symptom of a speech disorder: a sometimes overwhelming sense of isolation and loneliness. One

Victorian speech therapist described the 'habit of secrecy' of the person who stutters: 'of feeling himself cut off from his kindred; of brooding over his thoughts, of fancying himself under a mysterious curse'.[24] Francesca Martinez recalls the 'growing unhappiness and sense of isolation' of life with cerebral palsy, and those with aphasia talk – when they can – about being lost and lonely.

Isolation and loneliness are symptoms of the lived experience of many disabilities, but are peculiarly heightened by the very nature of a speech disorder. Given that speech is a tool for communication, then anything which hampers its efficacy separates an individual, even just a little, from the usual comforts of human discourse. My cousin Gilly finds herself being increasingly silenced in certain social situations. 'My projection is woeful and people talk over me all the time,' she says. There are certain types of interaction she struggles to participate in. 'If somebody is really aggressive or dogmatic, I can't take them on.'

While there are many different speech disorders, and immense variations within each, one thing above all unites them: the disfluent speaker has a fundamentally different relationship with language to fluent speakers. It is something that cannot quite be trusted. This is more than a matter of wrestling with an unreliable tool, like a cheap vacuum cleaner, because language is what ties human society together. No wonder, then, that feelings of alienation, of simultaneously observing and participating in any social situation, of never quite being in the moment of verbal communion, are so frequently described. This forced distrust of speech not only unites those who have a speech disorder but creates a gulf between them and everyone else.

The isolating qualities of a speech disorder are manifold. There is the isolation of humiliation, of having to forge one's own path through life, of feeling estranged from human communion. But there is something else too, although it is the hardest quality to describe: a feeling of detachment not only from others and from language, but from oneself. In psychology, the term *dissociation* describes the occasional but unsettling feeling of disconnectedness from yourself and the world around you. As a recurring condition, it is linked to childhood trauma and stress. When we consider the bullying that a child may experience because of a speech disorder, as well as the stress of continually needing to manage one's own speech with avoidance and substitution techniques, it is understandable why so many describe the symptoms of dissociation.

Darcey Steinke describes the out-of-body experience she had in speech therapy as a child:

My mind unfocused and I floated up, watching the skinny, pathetic girl in the sundress and tyre-tread sandals trying so desperately to find a little grace. Flash forward twenty-five years. After a plethora of speech therapy, my stutter was less disruptive; I moved through the world trying to pass as a fluent person, one unmarred by disability. Whenever I stuttered, I disassociated: that struggling human was not me.

But I think there is a bigger cause of dissociation in

speech disorders than just stress and social alienation. We use speech to describe our thoughts and feelings. When this faculty is disordered we find ourselves at one remove from the thoughts in our mind. The mental process of word substitution, for instance, is one of adjusting what I want to say to what I can say. It means looking at my own thoughts as an outsider. This might explain why I began to experience intense and disturbing attacks of dissociation when I was still a child, roughly around the same period I began to stutter. These attacks have never gone away although, like my stutter, they are less frequent and more easily managed. The two are forever connected in my mind and I see dissociation as an extreme expression of the internal distancing and editing I do with the thoughts and words I want to express.

While stuttering, vocal tics and dysarthria often develop from infancy over many years, the sudden onset of aphasia later in life is also explained in terms of intense dissociation. Some of these are captured in *Jumbly Words, and Rights Where Wrongs Should Be: The Experience of Aphasia from the Inside*: 'I live outside myself'; 'There was somebody else in my skin'; 'My mind sits on the fence'; 'I felt I was neither here nor there, just flotsam and jetsam'.[25] Understandably, such people also experience overwhelming depression, loneliness and anger.[26] One approach in treatment involves looking at aphasia through a grief paradigm – equating its impact to the loss of a loved one, although the person lost is oneself – moving through stages of denial, anger, bargaining, depression and acceptance. This can be far more than an emotional transition: an individual's personality itself can fundamentally change, as if unanchored and searching for some other mooring to drop in.

Speech disorders, then, are far more than conditions affecting the ability to say the words you want to say. They create moments of immense humiliation, estranging you from families and peers, unbalancing – often permanently – your sense of worth. They determine the paths you choose through life: the careers you embark on, the circles you move in, the associations you forge. They inform your performance in a social context: what you say, as well as when and how you say it. And, because speech disorders redefine your relationship with language, your very personality may change.

Often the experience of all these things feels like a closing in of horizons. Speech therapists find themselves supporting an individual through emotional crises as much as imparting techniques for the voice. I saw the same speech therapist, Willie Botterill, on and off from my pre-teens well into my twenties. On certain occasions, we barely spoke about my speech at all, but the anxiety I felt about it. Sometimes these enhanced periods of stuttering coincided with difficult periods in my life, like the death of a grandparent or the break-up of a relationship, so the boundary between speech therapy and counselling became negligible.

Many others reach far greater depths of despair than I ever imagined. American speech therapist Charles Van Riper described his many thoughts of suicide driven by his stutter, a condition he compared to being 'naked in a world full of steel knives'.[27] In the UK, one of the few charities that supports research into stuttering is the

Dominic Barker Trust, named after the twenty-six-year-old who tragically took his own life in 1994. It is a great sadness that every support organisation that deals with speech disorders can testify to those who have experienced despair or committed suicide.

This is why it is so important that we strive to understand speech disorders better. Not only as neurological conditions impacting speech, but as disorders of the mind, impacting mental health and life choices. And there is a third element too, which has emerged over the previous pages. The scale of psychological suffering an individual experiences is determined not so much by their inner resilience, but by the discrimination of the society in which they live. That sense of ever-narrowing horizons may be more than metaphorical but real, because humiliation, bullying and workplace prejudice undoubtedly eliminate opportunities for personal fulfilment. Like any social behaviour, the scale of such discrimination is not fixed, but varies both within and across different societies. Speech disorders, or at least the experience of them, are therefore shaped by culture. In this regard, they should be considered social disorders as much as neurological and psychological ones. This is both a cause for hope and concern. Social prejudice can get worse, but it can also, in the right conditions, get much, much better.

3

Talking Culture

In the late 1970s, the American linguist Daniel Everett began a life-long study of the languages of hunter-gatherer communities in the Amazon; in particular, the Pirahã people who live by the Maici river. The tribe, and the people who speak Pirahã, number less than a thousand. Besides the odd Portuguese word, they are monolingual. Their language is a closed circuit with its own unique structure and grammar: a precious rarity in a globalised world, where cultures seamlessly blend into one another. While wary of viewing contemporary hunter-gatherer communities as a paradigm of the forgotten origins of the West, Everett found it impossible not to draw insights into some of the ways language may have developed in the obscured past of our own civilisation.

In comparison to technologically advanced societies, Pirahã vocabulary and grammar is relatively narrow. Everett soon discovered how limited the role of verbal speech is in communication. 'You can recount jokes or lie, talk about the hunt, ask about the family or tell tall tales – all by whistling,' he wrote. 'In addition to whistle speech, the Pirahãs have hum speech, another form of communication that only uses pitch, loudness, and length, yet none the less communicates all the richness of normal human speech.'[1] Because industrialised societies pride themselves on their vast vocabularies (The Oxford English Dictionary Online, for instance, boasts over 600,000 words), they struggle to take notions like whistle and hum speech seriously, yet for Everett those modes of communication reveal a nuance and sophistication no less remarkable than our linguistic dexterity. In revering speech above all else, we may be missing out on other forms of communication. While we can express the abstract nuances of political ideology or religious belief in words, we struggle to express anything more than the most basic emotions of jauntiness or exasperation when it comes to whistling or humming.

Gradually, Everett became convinced of a simple idea that would nevertheless prove immensely divisive in the world of linguistics. His theory is an old, but unfashionable one: that language is simply a cultural tool aimed at communication and social cohesion. It is something human societies create, like any other tool, to serve a purpose. This may sound like common sense but it runs counter to some modern theories of language. In particular, it brought him head-to-head with his former teacher, the legendary linguist and political theorist Noam Chomsky.

Since the late 1950s, Chomsky has tried to prove that certain basic grammatical structures are innate to the human brain. As human societies develop they simply unfold this inner formula, or 'universal grammar', into reality rather than building it from scratch. While vocabularies need to be created, some of the grammatical structures they sit within are hard-wired into us. This is very different to a tool like the wheel or the hammer, neither of which existed as concepts in the human brain before they were invented. Everett argued that, after actually studying hunter-gatherer communities rather than speculating about them, there was no evidence of a universal grammar at all, nor any apparent benefit to be had if one did exist. By flying in the face of a sixty-year-old theory that has been institutionalised within universities, Everett brought the wrath of academia upon his head; he has been called a charlatan and a fraud by both Chomsky himself and his followers. The sheer level of vitriol in these attacks against what is, after all, a commonsensical theory of language only suggests the extent to which he has rattled them.

As a cultural tool, the Pirahã language evolved to suit the lifestyle and needs of its speakers. While it has a comparatively limited vocabulary and grammar, it is more diverse in its different types of speech. Whistle speech, for instance, is useful when out hunting: it enables people to communicate at a distance without shouting and startling the prey. In industrial societies, on the other hand, the complexity of a language evolves alongside scientific innovation with an ever greater prioritisation on uniformity and exactness. Theories of gravity, of relativity and quantum physics expand the limits of a language, creating a bedrock for further theories in the process. They would be impossible without a large vocabulary of words that are both specific but also abstract. We can't see gravity, for instance, but we all agree what it is.

But if speech and language are culturally determined, it raises the question of the extent to which speech disorders are too. In contrast to the speech of a fluent majority, the sound of somebody who stutters, emits vocal tics or has trouble articulating strikes us immediately. It sounds wrong and can even be unsettling. But what seems obvious may not be objectively so. This is the question I asked Everett when we met up, not in the Amazon, but in Liverpool where he was on an academic residency. 'In every Western society, the reactions to disfluency are fairly similar,' he told me:

> There's a real pressure to conform. Whereas I've noticed in some of the smaller societies there is more variation at the individual level in pronunciation, all sorts of variation you wouldn't find here. So I have encountered disfluency in Pirahã speakers. But nobody even comments about it or appears to notice it. I have seen people who had severe difficulty articulating consonants like p, t and k and instead would use a glottal stop [an explosive consonantal sound produced by obstructing airflow in the vocal tract]. And I've asked questions about it because I'm a linguist and I need to know what's going on and they almost get offended by the question.

While we are highly attuned to even the most barely perceptible disfluencies of speech, in Pirahã they are just characteristics of an individual's way of speaking. Even when an outsider introduces a Western 'speech disorder' into their language, the reaction is the same. 'One of my co-workers when I was a missionary had the worst stutter I ever heard,' Everett says. 'The Pirahã may very well have noticed there was a difference, but they never showed the slightest interest or concern in it.' Gradually, Everett began to experience it the same way. 'When I first met her, it was very hard for me not to act like there wasn't anything abnormal happening. I knew what she wanted to say, everyone knew what she wanted to say, and she was taking a while to get it out. But after I got to know her, I didn't even think about it any more because we spoke to each other all the time.'

Everett's observations about the lack of interest or even awareness of speech disorders in Pirahã highlights the extent to which they may be culturally determined. Disfluencies like stuttering may simply be idiosyncratic ways of speaking that are pathologised as medical problems within a particular social and cultural context.

I have described both the behavioural and psychological dimension of speech disorders, but this third – the cultural dimension – is the hardest to grasp. It is easy to understand the physiological notion that some people struggle to release their words in the effortless way others do. The idea that a person's psychological response antagonises this makes it a little more complicated. That culture might minimise or exaggerate the effect of a disorder, or even determine its existence entirely, seems to contradict the behavioural evidence. Surely a stutter or a tic is a neurological fact: it occurs, involuntarily, and we all recognise it when it does. But Daniel Everett is by no means a lone voice in questioning these assumptions.

In their essay 'Diversity Considerations in Speech and Language', two speech-language pathologists called Brian Goldstein and Ramonda Horton-Ikard argue that 'Speech and language acquisition does not occur in a vacuum but is mediated by the culture from which the child comes.

This environment is defined broadly and includes, but is not limited to, parents, siblings, extended family members, peers, teachers, and so on.' Speech is only considered defective 'if it deviates from the norms, expectations and definitions of his or her indigenous culture.'[2] In a small community like Pirahã, that deviation is simply everything that is not Pirahã. Indeed, the term they use to describe other languages translates as 'crooked head'[3] – something askew. But any variation within, by virtue of being Pirahã, is accepted. 'The Pirahã is like one big family,' Everett says, 'so I think that has something to do with the toleration of much wider degrees of individual diversity. There would be no reason to comment on your stutter, for instance, because I've known you all my life and that's no shock to me.'

In industrialised societies, you can spend all your life in the same small market town, let alone a big metropolis, and still not know everyone you encounter. In this context, small variations in speech and dialect become incredibly important: a way of denoting us and them. Sometimes we describe these differences as dialect or accent; sometimes as defectiveness. Historically, children in the United States who spoke African American Vernacular English (AAVE) were frequently diagnosed with a language disorder. This continued right up until the turn of the century when AAVE, also known as ebonics, began to be recognised as a systematic and rule-governed dialect in its own right. There is also evidence that the very nature of speech disorders varies across languages and cultures with physiological traits that are unique to each community.[4]

Within a society, attitudes towards speech disorders

may vary according to class. To take one example: Elizabeth Bowen, author of *The Death of the Heart* and *The Heat of the Day,* stuttered all her life. The circles in which she moved were distinctly upper class. Those she encountered socially found her stutter simply enhanced the eloquence of her speech. One friend described 'the stammering flow of her enthralling talk', another how her impediment gave 'an attractive touch of diffidence to her wide-ranging conversation'.[5] One British Council representative who booked her to give a talk in Zurich, described her speech impediment as 'endearing' in his official report. He concluded, 'She is a *most* successful lecturer with a *most* successful stammer.' These attitudes were typical of the British upper class at the time. Speech impediments suggested a certain unworldliness and rarified existence, like the stutter Evelyn Waugh gave to his 'aesthete par excellence' Anthony Blanche in *Brideshead Revisited,* in contrast to the smooth talk of the world of business.

And then there was the opinion of the wider population. In 1956, a BBC radio producer wrote an internal memo explaining why they had kept Elizabeth Bowen away from the airwaves, while embracing so many lesser-known writers. 'Elizabeth Bowen', it begins, 'is a stammerer. That is why we have never used her on such a big undertaking before we had tape.' Now, empowered by the innovation of pre-recording, their recommendation was to give her a try. 'We believe that she will be able to speak rather more fluently, however, if she is allowed to speak unscripted from notes and is allowed – if necessary – to rest during recording or to repeat difficult passages.'

Whatever permission that was required was soon granted and, from the insights of the BBC's Audience Research Department, we know how the general public reacted to Bowen's ensuing broadcasts. Despite the best efforts to re-record and edit out her trickier moments, listeners were irritated by Bowen's 'slow, jerky and hesitant delivery' and found her stutter 'painful to hear'.[6] It seems that what was endearing and scarcely disruptive for Bowen's upper-class associates was grating and distracting for the rest of the population.

The role of culture in defining and describing speech disorders is unsettling. The trouble a person has with stuttering, ticcing, using or articulating words seems innate, a physiological fact, even if symptoms are variable and exaggerated through the psychological experience of them. But by foregrounding the cultural dimension of speech disorders, we don't negate these other elements. We do, however, broaden and enrich our understanding of them.

This cultural dimension is in itself complex, although easily broken down into three main areas. The first is sociological: how a society, or elements within a society, react to any form of difference, turning what another society might consider negligible into an intractable problem. The second is linguistic: how a society's attitudes to language, and the labels it uses, enhance artificial constructs about good and bad speech. And the third (the subject of the next chapter) is culture in the more traditional sense: how the ideas and arts of a society work to include and exclude different ways of speaking.

Together, these three elements create an exaggerated

divide between those who are deemed fluent and those who are deemed disfluent or disordered in speech. The only way to debunk that binary perception is to expose the stereotypes and assumptions that support it. This is, in turn, the necessary first step to the rehabilitation and even appreciation of speech disorders within society.

Let's start with the sociological aspect. I have defined a 'speech disorder' as a consistent and obtrusive struggle in communicating the words you want to say. For the majority, that struggle is small and doesn't, in the case of the person with a slight stutter or early stage Parkinson's, prevent them from saying what they want to say. For others, as in severe aphasia or cerebral palsy, it is fundamental, becoming the defining or determining factor in how they communicate (sometimes an individual ceases to struggle, often because they choose to use alternative means of communication like Sign or Augmented and Alternative Communication devices, and when that happens we rarely think of them as having a speech disorder as such).

For some, the physical sensation of struggling with words may not be particularly negative. Any pain it causes, the additional labour in verbal performance it requires, or the impact on overall words spoken, may be small. The real difficulty emerges when that disorder is placed in a social context. This is almost always unavoidable because speech is, after all, a social tool. We can think of speech disorders, therefore, as mostly dormant conditions that are triggered when an individual comes into contact with others. It is only then that the worst symptoms – avoidance and concealment tactics, alienation and depression – are also activated. But unlike the physical symptoms, these psychological ones do not dissipate outside of social contact: they become all pervasive, haunting an individual's solitary hours. So while speech disorders are increasingly acknowledged and talked about as neurological in origin, their development and fruition is dependent on social factors. As we saw in the last chapter, the main determinant in a person's psychological response to the experience of a disorder is how family, schoolmates, teachers, co-workers and strangers react to it.

In his descriptions of the Pirahã, Daniel Everett offers the enticing vision of a society that simply ignores such disturbances in speech and communication. Unfortunately, this is far from the case with ours. From the schoolyard bullying of the kid with dysarthria to the infantilisation of the person with aphasia to the frequent discrimination in job interviews against the person with a stutter, the litany of acts of exclusion against those with speech disorders is endless. The trauma of these encounters encourages a secretive and self-censoring mentality in an individual who may retreat from society, or even attempt suicide to avoid further humiliation.

In some cases, social responses not only exacerbate a speech disorder, but transform its primary, or behavioural, symptoms. For people who stutter, problematic sounds and words shift and evolve according to the anxiety they cause. It is no coincidence, after all, that after years of stammering during school and even workplace roll-calls, the sounds of my own name became the hardest for me to utter. Similarly, when first describing the symptoms that would later become known as Tourette's syndrome in 1825, the physician Jean Itard concluded that the more his patient Madame de Dampierre was revolted by a word's 'grossness … the more she is tormented by the fear that she will utter them, and this preoccupation is precisely what puts them at the tip of her tongue where she can no longer control it'.[7]

'Rude tics are about social context,' says Jess Thom. 'Those I'm most frightened of are the ones that I think would most damage other people because I don't want to damage anyone else's confidence or self-esteem or sense of self.' As a result, she tries to curate what language she is exposed to, avoiding situations, television programmes or music that might highlight racial stereotypes or derogatory names. 'It seems bizarre on the face of it,' writes medical historian Howard Kushner, 'that something as rooted in culture as the utterance of inappropriate phrases or obscene words could be attached to organic disease … Even if neurobiology and biochemistry play an important role, a brain cannot "curse" without knowledge about what a culture views as a linguistic taboo.'[8]

The role of context is less vital in aphasia and dysarthria as they are generally linked to brain damage or neurodegenerative conditions, but many still report their speech varies according to how relaxed they are. Anxiety or humiliation are stressful emotions that temporarily worsen speech performance. My own stutter activates in certain social situations, such as public speaking, particularly if I am doing so off the cuff. But it also occurs, on the opposite end of the spectrum, in moments of great intimacy and emotional vulnerability. While I don't entirely share Wendell Johnson's view that a stutter is created by the ears of others rather than the mouth of a speaker, I think it is clear that the scale of a disorder is connected to the willingness, or lack of willingness, that a society has to accommodate it.

While speech disorders are considered neurological conditions in their medical context, in sociological terms they may be classified as a form of stigma. In his landmark 1963 book *Stigma: Notes on the Management of Spoiled Identity*, the sociologist Erving Goffman defined his subject as 'the situation of the individual who is disqualified from full social acceptance'. The word is Greek for a bodily sign like the branding of a slave, designed to show something unusual and bad about the bearer. 'We believe the person with a stigma is not quite human,' suggests Goffman. 'We tend to impute a wide range of imperfections on the basis of the original one.'[9] This is why we often suspect that those with speech disorders may have some sort of mental impairment or psychological trauma.

As Emperor of Rome in the first century AD, Tiberius Claudius Caesar was both man and deity. But not even his divine status could save him from the unpardonable offence of having a speech disorder. For centuries it has been assumed this was a stutter, but recent scholarship suggests that he had a form of cerebral palsy called Little's disease that may have prompted symptoms of stuttering as well as difficulty in articulating. Either way, he was considered the most ungodlike of living deities. According to Suetonius, his own mother considered him a 'monster of a man'.[10] Shortly after the Emperor's death, Seneca the Younger wrote a satire describing his apotheosis, or ascent to the heavens. In this work, Claudius's last words roughly translate as 'O no! I think I have shat all over myself.' He arrives at Olympus, but none of the gods can understand him because of his unintelligible speech.[11]

There is another side to the story though. Surviving documents suggest he could be effective at public speaking, even if his voice sounded a little odd.[12] Clearly there was a disjunction between the actual difficulty his speech disorder presented to communication and the way Roman society decided to interpret it. This, then, is the difference between a physiological condition and a stigma. The first is real; the latter a collective fiction which through critical mass takes on the appearance of reality.

While the neurological disturbances behind a speech disorder may be fundamentally the same across continents and epochs, a stigma is culturally dependent and therefore fluctuates. Not only across different societies, but also within them. Even in my lifetime attitudes to different types of speech in the United Kingdom have changed immensely. When I was born in 1975, minority dialects and accents were treated as stigma akin to disfluencies. In *The Presentation of Self in Everyday Life*, Goffman linked 'dialect and sub-standard speech' as negative character traits.[13] Geordie, Mancunian, stuttering and dysarthria all thrown in the same bucket. In the UK, the term 'Received Pronunciation' (RP) emerged in the 1920s to describe approved or 'normal' pronunciation. RP has been called at different times 'public school pronunciation', 'Oxford English', and more recently, 'Standard Southern British English'.[14] Because the BBC emerged at the same time as RP, a clipped, southern tone became the official lingua franca of the nation's broadcaster. Everything else was 'other' and implicitly inferior.

When my parents began their careers as journalists in the 1960s, they cultivated RP to get on in the world. For my father this was simply a question of exaggeration. His working-class father had already hacked away his cockney roots and catapulted his family to Home Counties splendour. When my sister and I found old recordings of dad doing interviews or reports we used to roll around laughing because he suddenly sounded like Prince Charles. Such reinvention in the United Kingdom was harder for my mother who was just off the boat from Australia, but her accent quickly found an inoffensive middle-ground somewhere between Sydney and London.

Forty years on, regional accents are celebrated in the media and in professional services (particularly call centres). In the BBC, where I work, RP has given way to a plethora of accents not only from across the UK, but the world (there is even a BBC News Pidgin service which, although considered a language rather than a dialect, is mostly comprehensible to the English speaker). There are many reasons for this change, but they all come down to the same thing: we no longer find it acceptable to openly discriminate on the basis of class or regionality.

For those with speech disorders, the dramatic and ongoing shift in attitude towards regional dialect is enviable but also a source of hope. But while there are signs of some change (I will describe some of these later), the stigma remains. Except as an occasional novelty, people do not stutter, tic or struggle to articulate on media outlets and certainly not call centres. It might be argued that this is because dialect is more intelligible than disordered speech to a fluent mainstream, but this is not the case. Most of us can recall situations when we've struggled to understand other dialects. On the other hand, speech disorders are unusual but generally intelligible, except when extremely advanced.

I have talked about two historical leaders, Claudius and George VI, whose difficulties with speech became a form of stigma. At the time of writing, Joe Biden is President of the United States. Biden has reminisced over the years about his childhood stutter. He has even implied his sense of political justice comes in part from the humiliation he experienced at school: his first act of protest was walking out of a classroom when a teacher mocked his struggle to say his surname. Now in his late-seventies, Biden's speech seems affected in different ways. During his campaign for the presidency, he sometimes appeared to forget words or say the wrong things. He once referred to himself as a 'gaffe machine'.[15] This led to frequent claims, mostly by rivals, that 'Sleepy Joe' suffers from the memory lapses symptomatic of mental decline and is not fit for office.

Around the summer of 2019, John Hendrickson, political editor of *The Atlantic*, became convinced that Biden's childhood stutter and late-life gaffes were not separate issues. Hendrickson, also a person who stutters, noticed a pattern to Biden's gaffes. They tended to happen around words with similar sounds, for instance. Often it was obvious that Biden hadn't actually forgotten what he wanted to say, but was trying to find other words with which to say it. Hendrickson concluded that Biden had never 'overcome' his stutter, but was in fact stuttering the whole time: blocking on words and using circumlocution to avoid them.[16]

Hendrickson's theory is that Biden, confronted with a choice to stutter openly or feign forgetfulness, was choosing the latter. The cost of this is immense. 'At an August town hall', Hendrickson writes, 'Biden briefly blocked on Obama, before subbing in 'my boss'. The headlines afterwards? 'Biden forgets Obama's name.' Hendrickson suggests Biden's reluctance to admit to stuttering was partly because of the story he was trying to tell about himself: his stutter was one of those triumphs over adversity that presidential candidates like to talk about. But there is another reason too. Political campaigns are carefully choreographed events with candidates and their advisors shaping the message they want to present to the world. If Hendrickson's theory is true (and I find it compelling), then it is possible Biden felt more comfortable suffering regular accusations of mental decline than admitting to stuttering. Whether this decision occurred at a conscious or unconscious level is impossible to tell as concealing a stutter becomes second nature for those, like me, who can. Either way, Biden's story is another reminder that while Claudius and George VI are figures of bygone times, the issues they faced haven't gone away in the slightest.

So why has the stigma of speech disorders proved so intractable? Goffman suggested one theory. While support groups and activist movements have long existed for almost every type of social stigma (whether alcoholics, ex-convicts, the aged, the obese, the physically handi-capped), he claimed, 'there are speech defectives whose peculiarity apparently discourages any group formation whatsoever.' Goffman found this perplexing considering how disabling such conditions could be. Since isolation and secretiveness are key psychological symptoms of speech disorders, this is understandable. But I believe it also emerges from a tendency to look at speech disorders in medicalised silos ('Tourette's syndrome', 'aphasia', 'dysarthria') rather than as different expressions of the same discriminating social framework.

For most of my life, I embodied the 'peculiarity' that Goffman writes about. I refused to talk about my speech except in safe institutionalised spaces like the Michael Palin Centre for Stammering Children. I maintained this silence even when I encountered other people who stut-tered. On one occasion, in a restaurant in Newcastle, I stared blankly at a clearly distressed waiter who stuttered throughout our exchanges when I'm sure one kind word of solidarity would have alleviated his suffering. And I still recoil in self-disgust at my inexplicable laughter in the face of a stuttering German student in a youth hostel in Rome some twenty years ago. I can only explain it as a moment of irrational hysteria, as if coming upon a dop-pelgänger of myself, as I had encountered so few people who overtly stuttered at that stage in my life.

With so little empathy even for other people who stutter, the idea that I had anything in common with other speech disorders never even occurred to me. Sometime in the late 1980s there was an attempt to set up a National Speech Day to raise awareness of speech disorders. My mother and I joined a press event in Hyde Park where I was photographed alongside children with other conditions and speech disorders. I remember looking down the line and thinking, 'what am I doing here?' In keeping with Goffman's theory, National Speech Day soon disappeared. The only evidence I have of it is an old newspaper photograph of a minor celebrity advocate, a row of children and my strained smile. It is only in the last five years that I have started talking to people with the whole spectrum of speech disorders and realised how much we have in common.

There is one other reason, but perhaps the biggest, why the stigma of speech disorders remains so entrenched. It is not to do with such disorders themselves but of what they are defined against: the notion of fluency. Although an ongoing process, dismantling the stigma of dialect has been enabled by a growing awareness that RP is a sacred cow worth killing. For while 'public-school pronunciation', 'Oxford English' and 'Standard Southern British English' may dominate our culture still, they are also associated with privilege, elitism and inequality. But how does one go about denigrating a notion so apparently virtuous as 'fluency'? It may not be possible to do so, yet without it speech disorders will always be considered a sub-standard expression of language.

How a society thinks about its languages and the labels it gives to its different usages enhance ideas of good and bad speech. The very term 'speech disorder', after all, determines from the outset how we are to think about stuttering, aphasia, dysarthria and vocal tics. As a child, just knowing I had a 'stutter' made me extremely conscious of the blockages and repetitions in my speech. Similarly, many of the social prejudices about speech disorders are determined the moment they are labelled as conditions, and those prejudices, like the label itself, inform and exacerbate both the psychological and behavioural traits of those conditions.

Our ability to even think about this issue is hampered from the outset: *disfluency, disorder, abnormal* and *impediment*. These terms are radically different, even oppositional, to another set: *fluency, order, normal, unimpeded*. Linguistically, a thick line implicitly separates normal speakers from those with speech disorders, often through the use of binary, negating prefixes like *dis-* or *ab-*. The moment we start talking about disfluency or disorder, we are already assuming a degree of defectiveness; an influence that is difficult to evade. This is a trick of linguistics rather than a reflection of reality, for once we label something in a negative way it determines how we view it. (For instance, if somebody tells me that a particular person is 'immoral', it will colour my perception of them until I am convinced otherwise.) So if we strip away these binary positive and negative terms, how much difference is there really between 'fluent' and 'disfluent' speech?

When we encounter a person with a speech disorder, our brains try to make sense of a difference that may be extremely pronounced or even quite subtle. Labelling is key to this process. If I meet someone whose eyes are different colours (a condition called heterochromia iridum), it may take me a moment to realise what is unsettling me, but once I identify the difference I feel more relaxed. Likewise, if somebody's speech is scattered with disfluencies, it will perplex me until the point that I determine that they have a stutter; or if interspersed with vocal tics, I may start to suspect Tourette's syndrome. Such labelling doesn't require knowledge of medical terms. After all, many people who have dysarthria aren't even aware of this term. But it requires knowledge or experience of a type of vocal dysfunction. Sometimes labelling may just be a suspicion that one stranger who speaks slowly and uncertainly might have had a stroke, while another who is unsteady and has unusual articulation may be in the early stage of Parkinson's disease or motor neurone disease.

Whether named or not, this process of labelling is important for us, as all those individual disruptions in the way a person talks suddenly become expressions of a single cause or characteristic. It is often accompanied by a sense of relief. We don't like uncertainty, particularly when dealing with strangers. 'People are so used to whatever their concept of normal human interaction is that when someone different comes along they are thrown,' says Lee Ridley, a stand-up comic with cerebral palsy. 'They're not used to a bloke who can't communicate in the usual way. They're not getting the well-versed social signs through body language and facial expression, so they're not sure how they're doing. Am I saying the right things? Am I keeping eye contact in the right way?'[17]

The problem is that once we label a person in this way, we burden them with all the preconceptions and prejudices we may have about a condition. Labels, like all forms of categorisation, are useful illusions that help us make sense of the world, but they also lead us to attribute general qualities that may not be relevant to particular individuals. Our assumption is that a *dis*order of speech is a breakdown of function. This leads to discrimination in the workplace and other social environments.

Take, for instance, the common perception that word flow – and, therefore, productivity itself – is slower or reduced in a person with a speech disorder. While this is true in some cases, it is by no means a universal trait. People who stutter often compensate by speaking quickly between stuttering events. The rhythm of their speech may swing between extremes but they don't necessarily get less words out. While for people with Tourette's syndrome, vocal tics often fill up the natural pauses that arise in speech rather than creating them. Even if speech is rendered slower, it may not be less effective. For Jamie Beddard, dysarthria makes him more economic in his choice and use of words, but not in his overall ability to communicate. Rather than letting our labels and prejudices determine the capability and limits of a person's speech, we would do far better to simply familiarise ourselves with their unique speech patterns just as we unconsciously do with all the 'fluent' speakers we encounter.

If certain people have speech disorders, impediments or disfluencies, then those who don't – a sizeable majority – are implicitly 'fluent'. Studies of speech disorders rarely attempt to define what we mean by this, dragging the reader into an oppressive and sometimes airless space in which the notion of a 'disorder', and the gulf separating it from normal speech, is assumed from the outset. Yet it is impossible to truly make sense of such conditions unless we are clear about what it is they are apparently failing to perform. Just as we need to define disfluency, we also need to define fluency.

'To be a fluent speaker of a language means to be able to enter any conversation in ways that are seen as appropriate and not disruptive,' writes linguist Alessandro Duranti.[18] According to David Crystal, fluency is the 'ability to communicate easily, rapidly, and continuously'.[19] The word itself derives from the Latin 'fluere': to flow. The fluent tongue is like a river, gathering momentum and speed as it moves from a mountain spring into a great estuary opening out to sea. But despite images of flowing rivers and unimpeded tongues, 'fluency' is far more complicated than at first appears. It's a quality that applies to a lot of people and yet there is a huge amount of variation between them. A great orator like Barack Obama as well as the mumbling teenager in the upstairs bedroom; the clipped tone of the Queen of England as well as the broadest dialects of our remotest regions: all these may be considered fluent.

While there are those who have a quiet, or not so quiet, confidence in their speaking ability, just as many of us feel tongue-tied, lacking the ease and naturalness of speech that fluency implies. Rather than setting up fluency and disfluency as binary opposites, it would be more accurate to think of fluency as a spectrum in which people move back and forth over the course of their lives. Even this is of only limited value for it might capture the sound of somebody's speech but not how they feel about it. There are fluent speakers who experience as much anxiety about their voices as those with diagnosed speech disorders. According to a YouGov poll from 2014, fear of public speaking, or *glossophobia,* is the third biggest phobia in the UK after heights and snakes,[20] fuelling an industry of voice coaches, workshops and self-help books.

The etymology and ideal of 'fluency' suggests an ease and rapidity of verbal performance, but what does it sound like in reality? In recent years, projects like CANCODE (Cambridge and Nottingham Corpus of Discourse in English) have recorded millions of words of everyday conversation. In transcribing conversation, researchers develop a grammar of symbols to denote the breakages in speech that don't occur in the written word. In the case of CANCODE, '+' marks the point where one speaker's utterance is interrupted by another. A common sign is '=', which shows when a speaker has either changed their course in mid-sentence or in the middle of an individual word, as in the following exchange:

Speaker 2: Yeah. I wonder now about people who go into the army these days like m= I had a friend erm who went who was a good friend of mine when I was at school you know+

Speaker 1: Mm.

Speaker 2: +up till sixteen. After that we lost touch a bit but he only lives round the corner from me now. But he went erm on a army course recently and he did really well and he got to the last thing at Sandhurst where you would go+

Speaker 1: Yeah. [laughs]

Speaker 2: +in as an officer+

Speaker 1: Yeah.

Speaker 2: +and he did the thing there and he fa= he fell at the last hurdle. And

Speaker 1: What he he, you're talking metaphorically here?

Speaker 2: and and er fitness trainings and did really well and now he's gotta go in as a, as a yu= you know as a regular.[21]

Struggling to keep up? What such exchanges repeatedly demonstrate is how disfluent 'fluent' speech really is, with persistently awful grammar, verbal stumbles, torturous and rather inarticulate attempts to express the ideas in our minds. 'The voice is supposed to be suffused with spirit,' writes linguist Steven Connor. 'But the voice is not always quite itself … it is full of poltergeists, noisy, paltering parasites and hangers-on, mouth-friends, vapours and minute jacks.'[22] Those qualities we associate with everyday fluency, an uninterrupted ease and rapidity of delivery, are extremely rare. We are, in fact, all disfluent, but in such a variety of ways that each individual slip is scarcely noticeable. It is only when one type of disfluency eclipses all others, like a vocal tic or the block of a stutter, that we describe someone as having a speech disorder.

All this is well known to speech-language therapists. In defining the traits of stuttering, for instance, the important distinction is not between disfluency and fluency but between abnormal and everyday disfluency. According to *The Handbook of Language and Speech Disorders*, disfluencies like *interjections* ('he went-um-home'), *word repetitions* ('he-he-he went home'), *phrase repetitions* ('he went-he went home'), *revisions* ('he went-he ran home'), *incomplete phrases* (he went …'), and *broken words* (he we-[pause]-nt home) are all part of fluent speech.[23] None of these should be considered symptomatic of stuttering, which is identifiable by part-word repetitions, some forms of word repetition, prolonged sounds and tense pauses. Add it up, and there are actually more types of disfluency in everyday fluency than in stuttering. Fluency, like disfluency, is an illusion. Once you accept this, it is difficult to hear human speech in quite the same way, but the repercussions are only positive. There is no need for anyone to feel overwhelming anxiety about their speech because there is no perfect standard to aspire to, while any discrimination against those with speech disorders is hypocritical.

The discovery that 'normal' speech can be just as disfluent as many speech disorders comes as a shock. One reason for this lies in the processing of the brain. In order to make sense of the conversations we participate in, it is necessary at an unconscious level to strip away all the non-essential information – the 'poltergeists, noisy, paltering parasites and hangers-on' – scattered throughout our speech. We only notice such disruptions if they have a pathological regularity that repeatedly draws our attention. The human brain looks for patterns in the material world and a speech disorder is just one such pattern. What is in reality a sliding scale is given a binary separation in our minds: normal, and therefore unnoticeable, disfluency; or abnormal, and therefore disruptive, disorders.

While ignoring the ums and ahs, the blockages and repetitions, we also over-emphasise the role of speech in communication. While we fixate on (and often regret) the things we say, sociological studies repeatedly show that much, even the majority, of human communication is nonverbal, based on body language and tone of voice as well as words.[24] Yet it is the words themselves we hang onto, perhaps because they seem more tangible and less ambiguous. There are, after all, many dictionaries that tell us exactly what they mean, while interpreting atmosphere and gesture is more often a question of instinct. But we also know that if somebody is hurt by our behaviour, it is more likely to be because we didn't look at them or turned our bodies away rather than any specific comment. And this is why we, in turn, are sometimes relieved or triumphant when somebody we have long suspected doesn't like us says something that actually confirms it.

If words aren't so important this should mean that speech disorders scarcely present a problem to the ease and rapidity of communication, but the opposite is true. In the 1970s, the psychologist Albert Mehrabian famously devised a rule which argued that our emotional response (liking, disliking or indifference) to a person is determined mostly by their body language (55 per cent), then their tone of voice (38 per cent), with comparatively little depending on the words they say (7 per cent).[25] Unconsciously, it seems, we are far less trusting in the power of language than we profess: and wisely too, bearing in mind what we know through speech corpuses like CANCODE. But it also explains why the immediate reaction to somebody with a speech disorder can be so negative. Since such conditions affect not only the tone of voice, but are often accompanied by further physical contortions in the face or body, listeners are more likely to notice them. They react far more to the behavioural signs of the disorder, allowing it to influence and even determine their emotional response to that individual, than anything they might have to say – and that, in turn, interferes with the process of communication.

The illusion of fluency and disfluency is partly down to human nature. It is created by our brains which prejudice in favour of what is tangible (words rather than gestures) and by our memory which simplifies the mess of conversation. Not only do we edit conversations as and after they occur, ignoring recurring and unobtrusive slippages of the tongue, but we overemphasise the role of speech in all interpersonal exchanges. It is hardly surprising then that many societies treat speech disorders as grand human failings that compromise the preciousness of communication.

But this illusion is also cultural: the reverence for fluency varies across human societies. In the post-industrial societies of the twenty-first century, it is considered far more than an enabler of communication, but the key to all success; the glue that ties society together; the essence of being human. This false conception has evolved through our 'culture', in the traditional sense of the term, shaped by philosophy, politics and aesthetics. As we are about to discover, our society defines disorders not only against fluency but an extreme version of it: a *hyper-fluency* that sits on the other end of a spectrum to disfluency, and is as far from everyday fluency as that is from disfluency.

4

The Tyranny of Fluency

In February 2002, a forty-five-year-old British entrepreneur stepped onto the stage of a lecture theatre in California and made an announcement that would change the course of his life. The event was an annual conference devoted to speakers from the worlds of technology, entertainment and design. It had been running for eighteen years but was on its last legs: its peak seemed past, its founder was retiring, attendance was down to only seventy. In theory, it had little power to make or break anyone's career except this man had made the dubious decision to buy it. TED was his gamble and he needed all the people in the room to back him.

Chris Anderson's vision was to take the content of TED – which, after all, had a track record of attracting some superb speakers – and put it on the internet. Making videos is straightforward enough, doing it in a way that people want to watch them remotely is far harder. In selling TED as a brand, Anderson had to sell an aspirational idea of what a lecture can be. His breakthrough was to recast the public speaker as hero.

This emerged through the aesthetics of what was soon recognisable as the TED style. There is the speaker's platform: empty, uncluttered, spotlit in the centre and fading away into darkness. There is the audience itself, revealed in dimly lit, wide shots that communicate their admiration but prevent any distraction from the solitary character on stage. There is the posture of the speaker, liberated of lectern and microphone through the means of a headset to stand alone, vulnerable and defiant before the crowd. But most of all, there are the speaker's words. These are talks that are full of humility, revelation and human stories. They are rehearsed over and over again, timed to the second, resulting in sometimes breathtaking performances of charismatic verbal fluency.

While some TED talks are more successful than others, there is a set of principles that runs through them all. Anyone can learn these simply by watching the content or reading the publications, official and unofficial, that reveal TED's 'secrets'. All emphasise the same things: the importance of authenticity and vulnerability, of telling stories rather than giving facts, of delivering jaw-dropping revelations, as well as what to wear and how much to rehearse. Most of all, the importance of eighteen minutes, which Anderson described as 'long enough to be serious and short enough to hold people's attention'.[1]

'Great speakers find a way of making an early connection with their audience,' Anderson writes in his book *TED Talks*. He takes as an example a talk by a successful health psychologist:

Take a look at the first few moments of Kelly McGonigal's TED Talk on the upside of stress. 'I have a confession to make.' [she pauses, turns, drops hands,

gives a little smile] 'But first, I want YOU to make a
little confession to me.' [walks forward] 'In the past
year' [looks around intently from face to face] 'I want
you to just raise your hand if you've experienced
relatively little stress. Anyone?' [an enigmatic smile,
which a few moments later turns into a million-dollar
smile]. There is instant audience connection there.

For me, the most striking thing about this transcrip-
tion isn't necessarily how effective Kelly McGonigal's talk
is, but how mannered and choreographed it is compared
to everyday speech and communication. The term Anderson uses to describe this is 'presentation literacy': the
art of 'unlocking empathy, stirring excitement, sharing
knowledge and insights, and promoting a shared dream'.
Anyone who worries about their ability in this regard is
right to do so. 'Presentation literacy isn't an optional extra
for the few,' he writes. 'It's a core skill for the twenty-first
century.' The success of TED suggests that he is right.

Within a few years, Anderson had turned TED around
from a small conference of waning influence to a global
phenomenon. The high contrast, visual simplicity of
the content – a lone individual against a dark backdrop,
often with hands thrusting out before them like a twenty-
first-century prophet – worked well for the early years of
internet streaming. Even pixellated, its power carried.
Thinkers, academics, entrepreneurs and activists, who had
struggled for mainstream recognition, became celebrated
overnight with immediate impact on book sales and busi-
nesses. Ken Robinson, Amy Cuddy and Simon Sinek:
brilliant but formerly low-profile intellectuals whose talks

have now been seen by over forty million people around the world. This in turn attracted major celebrities like Bill Clinton, Bono and Bill Gates. Seats at the conference itself became hot tickets and now cost thousands of dollars – just to see people talking. Before long, TED franchised the brand. According to their website, there are TEDx conferences in over 130 countries a year, with an average of eight events taking place every day: from Iran to India as well as America and the UK.

The global triumph of TED is symptomatic of a trans-formational moment in our relationship with a particular linguistic register. Neither Anderson's term 'presentation literacy' nor the more old fashioned 'public speaking' seem to quite describe it. While there are those individuals who become a different personality on stage, there are also those whose heightened fluency transcends both private and public discourse. They speak with extraordinary eloquence and expressiveness, both personable and memorable, seemingly without the everyday disfluencies of fluent speech. In this regard, their speech is as removed from everyday fluency as disfluency is supposed to be – and it explains in part why so many perfectly 'fluent' people feel such a sense of inadequacy. Not so much fluency, therefore, as hyper-fluency: speech without the ums and ahs.

The pressure to be hyper-fluent is something many of us know, fear, and are unable to avoid: not only job interviews and PowerPoint presentations, but also (less regularly) stage-of-life addresses, like those at weddings and funerals. It pursues us into seemingly innocuous social scenarios. The party piece and pub anecdote are second nature to some, but sources of anxiety and feelings of inadequacy for the many who struggle with them. We work hard at improving our performance in these arenas. The success of organisations like TED, as well as a large industry of voice coaches, public speaking gurus and self-help books, testifies to this. But how desirable is hyper-fluency really?

In *Quiet: The Power of Introverts in a World That Can't Stop Talking,* Susan Cain argues that our culture is dominated by what she calls an 'Extrovert Ideal'. Cain is convinced that the power of the introvert is lost in our culture and the Extrovert Ideal dominates everything we do: not just our professional lives, but our social and even domestic selves as well.

> Talkative people, for example, are rated as smarter, better-looking, more interesting and more desirable as friends. Velocity of speech counts as well as volume: we rank fast talkers as more competent and likeable than slow ones. The same dynamics apply in groups, where research shows that the voluble are considered smarter than the reticent – even though there's zero correlation between the gift of the gab and good ideas.[2]

While volubility, the quality of talking fluently, is identified as a common characteristic of the extrovert, the absence of it doesn't always imply introversion. Many with speech disorders struggle to get their words out irrespective of where they sit on the introvert–extrovert spectrum. But I do think we can apply the same scepticism about the supposed superiority of extroversion to hyper-fluency because of the considerable overlap between both qualities.

Undoubtedly, hyper-fluency can be a powerful and influential means of communication, as the finest TED talks show. But it has inherent risks and dangers too. Even when used with the best intentions, there is a gravitational pull to glibness and simplification, 'style over content' as the saying goes, that seems astonishing in the moment, but considered afterwards often seems nothing more than the obvious said well. While there is a popular misconception that the best ideas are simple (a statement which lends itself conveniently to a culture increasingly built around the power of short-form content), there is nothing simple about the Theory of Relativity: even Einstein struggled to make sense of it. Often in rendering complex ideas palatable, we strip them of their potency precisely by making them simple.

Hyper-fluent speakers can also be charlatans, con-artists and corruptors, exercising their skill for personal gain or just drunk on their own power. During the 1930s, the German linguist Victor Klemperer charted the way the Nazis secured their ascent through the manipulation of language.[3] The political success of Adolf Hitler and Joseph Goebbels lay in their oratory skills first and foremost, honed through years of public meetings and rallies. They consciously developed a technique that controlled a crowd through the repeated use of buzzwords, euphemisms and outright lies, but all under a veneer of supposed common sense and heartfelt passion. Klemperer provides many examples revealing just how elaborate this was.

The words *artfremd* (alien to the species), *ewig* (eternal, as in 'the eternal Reich'), and *volk* (people) were repeated endlessly. The word for 'murder' was replaced by 'special treatment', 'mass-murder' by 'final solution', while the war was always 'imposed' on a peace-loving Führer.

Less harmful, but worrying in a different way, is the way hyper-fluency can conceal an absence of content: verbal performances that are dazzling in the moment, but leave nothing behind. Our politics is polluted by a fluency prejudice which means leaders are often chosen by their resemblance to after dinner speakers rather than on the length and quality of their experience in public service. Meanwhile, social media is the making of multi-millionaire teenage YouTubers, who have a knack for talking compellingly about make-up or gaming with often little more than a bedroom wall behind them.

One of the most successful is Logan Paul, who began posting videos from his family home in Ohio when he was only ten years old. In his world, and that of thousands like him, hyper-fluency is not reserved for rare public speaking occasions, but is continuous: addressing an audience while in bed, on the way to school, out with friends. By the time he was twenty, Paul's videos had over 300 million views, and he was loved by kids for his banter and shameless boasting about his new-found wealth. The banter briefly ran out after he filmed and posted a video of a dead body in Mount Fuji's 'suicide forest'. Growing up on YouTube meant he was unable to separate the tragedy of somebody's dead son from clickbait. Over the real footage of the body he even pasted sentimental piano music to enhance its emotional power. Paul's response was no less bizarre. Rather than empathising with the victim or the victim's family, he said, 'this is a first for me,' as if the significance of the suicide lay in his experience of it.

Another problem with hyper-fluency is the way it both raises and narrows public perception of what qualifies as good communication. We come to think that there is only one acceptable style of speaking in public, which most of us aren't particularly proficient in. What gets lost in the mix is a diversity of public speaking styles: rambling talks which don't have an obvious message; uncharismatic talks by people too shy to look at the audience; talks which are just weird stories that don't obviously connect; talks which don't tell any stories at all but just give you facts; talks which haven't been learned by heart but sound like good prose when read from the page; talks which are very short or very long; talks by people who speak differently to others. All of which have a place in our culture, but of which our shortened attention spans are increasingly intolerant.

Marina Abramović is a world famous performance artist who has extolled the power of nonverbal communication through her work. She is perhaps best known for a show called *The Artist is Present* which ran for three months in 2011 at the Museum of Modern Art in New York. The concept was simple: Abramović sat in the same chair eight hours a day for three months. Visitors to the museum were able to sit opposite her and commune through silence and eye contact. It's not an immediately promising idea, but the response was extraordinary. Something happened in those silent exchanges. People claimed it changed their lives; many wept. Before it closed, people queued round the block overnight to seize the last opportunity for a silent encounter with Abramović.

Because of the success of the show, TED approached her to give a talk. The result is an odd marriage: an artist famous for advocating nonverbal communication restrained within an inflexible speech format. She looks uncomfortable. 'TED was really frustrating,' Abramović tells me when we speak on the phone. 'They time you. They wanted me to repeat and to verbalise things which I hated so much. I went crazy. I can't repeat things twice. It's just against me.' The difference between Abramović as a public speaker and as a nonverbal communicator is palpable in the videos of her TED talk and *The Artist is Present*. The latter have by far the greatest power. Hyperfluency is a form of communication but by no means the only one, nor the best.

There is one final reason why I think we need to resist simply accepting hyper-fluency as an imperative for twenty-first-century life. As well as narrowing the diversity of communication styles, inclining to glibness and, in the wrong hands, deception, it is also exclusive. We are told that, with a bit of practice, anyone can speak well. But as much as we're told the potential is within us all, many of us have communication styles of one sort or another – whether speech impediments or introverted personality traits – that make the pithy, charismatic talk as out of reach as the four-minute mile.

The truth is that some people are better predisposed to hyper-fluency and 'presentation literacy' than others. So when we're told it is a core skill for life, we are right to worry, just as we would if told that a 'high IQ' or 'toned body' isn't an optional extra for the few. It is just one linguistic register in a spectrum that includes endless manifestations of everyday fluency as well as diagnosable disfluencies. By elevating it to an imperative we risk what has in fact happened: widespread social and economic discrimination against those who speak differently.

I believe that it is only when we challenge the supremacy of hyper-fluency in our society that we also challenge the discrimination against other ways of speaking, including speech disorders. Doing so will not be easy, yet we can learn from the discrepancies between cultures. 'Westerners are typically uncomfortable with silence, which they find embarrassing and awkward,' write Adler, Rodman and du Pré in *Understanding Human Communication*. 'On the other hand, Asian cultures tend to perceive talk quite differently … Japanese and Chinese people more often believe that remaining quiet is the proper state when there is nothing to be said. To Asians, a talkative person is often considered a show-off or a fake.'[4]

In *Far From the Tree,* Andrew Solomon describes visiting a small village in northern Bali where a congenital form of deafness affects around 2 per cent of the population. Everyone there can use the unique sign language, called Kata Kolok, that they have developed. But what eventually struck Solomon wasn't the remarkable phenomenon of a community that was bilingual in spoken and sign language, but how secondary both were to community cohesion:

> For educated Westerners, intimacy requires the mutual knowledge achieved as language unlocks the

secrets of two minds. But for some people the self is expressed largely in the preparation of food and the ministrations of erotic passion and shared labour, and for such people the meaning embedded in words is a garnish to love rather than its conduit. I had come into a society in which, for the hearing and the deaf, language was not the primary medium through which to negotiate the world.'[5]

Even within the West, there are unique traditions that revere non-verbal communication. In many Native American communities, silence is considered the appropriate way to express deep emotions like grief and anger:[6] hence the stereotype of the silent Apache in many Western films. And in Europe, social order was maintained throughout the Dark Ages largely due to the network of Christian monasteries that reached across the continent, generally founded on strict rules that regulated speech to a minimum. When the writer Patrick Leigh Fermor began staying at monasteries in the 1950s, he found that 'the desire for talk, movement and nervous expression that I had transported from Paris found, in this silent place, no response or foil, evoke no single echo'. Instead, he needed less sleep and found himself with 'nineteen hours a day of absolute and god-like freedom'.[7] The difference is that these Western traditions of silence have now been enclosed within reservations or small compounds, forgotten amid the noise of a culture addicted to chatter.

These examples don't negate the fact that hyperfluency can be useful in 'unlocking empathy, stirring excitement, sharing knowledge and insights,' as Chris

Anderson describes it, but they do remind us that there are other approaches. While we consider hyper-fluent speech a virtue, whether in the formal context of a presentation or the storytelling of the pub wag, we do well to remember that our perception is culturally determined and therefore relative, just like our attitude to speech disorders, rather than a reflection of something objectively good.

If hyper-fluency is cultural rather than innate to human nature, the question is how it developed and what purpose it serves. In *TED Talks*, Anderson offers an answer. While recognising 'presentation literacy' as a particularly twenty-first-century imperative, he suggests it is a modern embodiment of rhetoric, or the art of persuasion, drawing an evolutionary line between the assemblies and market places of the classical world and the TED URL. There is some truth in this. While first described in Mesopotamian and Egyptian texts, rhetoric emerged as a major discipline in Greece around 500 BC, inseparable from the political notion of democracy that it both informed and was shaped by.

Democracy is a politics of consensus and debate (often frustratingly prolonged) rather than the individual prerogative of a tyrant, in which a command can be brief and irrational, but effective. The uses of rhetoric, therefore, extended beyond lawgivers and a priest class to all those who participate in some way in the city life, or *polis*, of Athens. By the time Aristotle came to write *The Art of Rhetoric* in the fourth century BC, it had evolved into a highly codified and ritualistic practice: a sort of martial art, or Tongue Fu, of the voice.

In Rome, which saw itself as both conqueror and inheritor of Greek civilisation, rhetoric became even more respected. Roman Law, which contains the origins of our legal system, is based on spoken testimony and rhetorical advocacy rather than the rough justice of many non-democratic communities. Both Cicero and Quintilian, the great theorists of Roman rhetoric, began their careers as lawyers, and developed theories that position rhetoric more holistically as the external expression of a virtuous and enquiring individual rather than a self-contained discipline. In this regard, rhetoric gradually became a social signifier, evidence of a distinguished background and good education, as much as a pragmatic tool for persuasion.

'In the world of the Roman aristocracy,' writes classicist Christian Laes, 'achievement in oratory was as glorious as success on the battlefield.'[8] This is the beginning of a subtle difference between rhetoric and hyper-fluency, for where rhetoric has a purpose, hyper-fluency is a quality that an individual may reveal during even the most banal and purposeless of exchanges. A hyper-fluent individual, rather than a merely fluent one, is somebody clearly well-bred and intelligent; somebody you can trust and do business with, while hoping to gain access to the rarified social circles they have access to.

Rhetoric is not the only legacy from the classical world informing the fluency-prejudice of the twenty-first century. Just as important is that of another group of wordsmiths who had a difficult, sometimes outright hostile, relationship with those who practised the art of rhetoric. While we tend to think of philosophy today as a solitary and silent practice (thanks in part to iconic images like Rodin's *Thinker* who is locked in an intense and mute reverie with mouth literally stoppered by his own fist), for most of our history it has been an intensely social and conversational practice. In Ancient Greece, philosophy was dominated by 'schools': groups of pupils and peers who gathered around revered thinkers. The term signified 'love of wisdom'. Its focus extended beyond the (often obscure) riddles of existence we tend to associate it with today, and included natural science, physics, astronomy, mathematics. In short, everything one might consider as wisdom – until the drive to specialisation of relatively recent times.

Philosophy depended, then as now, on the use of words. The movement of the heavens, the atomic structures of invisible matter, the moral quandaries of daily life: all this could be explained, if only we tried hard enough, through speech alone. One early philosopher, Heraclitus, even elevated the term for word or speech, *logos,* to mean knowledge and cosmic order itself, as if human language contained all of natural law within its structures.

Socrates, perhaps the most famous of all philosophers, preferred dialogue and collective reasoning, the so-called Socratic Method, over abstract thought or demonstration as the means to all knowledge. He never wrote anything down and we depend on the writings of his pupils, particularly Plato, to know his teachings. Plato's works are all dramatic dialogues, in which he presents his master Socrates and his friends uncovering the secrets of the universe by simply talking to one another. Their speech is effortlessly fluent and idealised, unhampered by any sort of disfluency, whether the hesitations or repetitions of everyday speech or the stumbles, tics and distortions of speech disorders.

Together, the classical traditions of rhetoric and philosophy form the origins of modern hyper-fluency. They remind us that it is a way of speaking that has as much to do with class and verbal gymnastics as well as communication. As a result, the Greeks and Romans came to see fluency as synonymous with civilisation itself. Those who weren't civilised were barbarians. The Greek word 'barbaros' is related to the Sanskrit word 'barbara', which referred to the speech-defective and fool as much as the uncivilised outsider. This in turn informed the Latin word for a speech defect: *balbus*.

Throughout the classical world, there is a deliberate association of citizens who had speech disorders with the barbarians beyond the frontier. This is reflected in the broader attitude to speech disorders which is marked by lack of interest and occasional contempt. Christian Laes has found that in all the documents of the classical world there are only fourteen instances where it is suggested a particular individual has a speech disorder. In each case it is often hard to tell what type of disorder they had because their terms were so vague, in itself a sign of indifference. *Balbus* was a catch-all word for the person who stutters, lisps or is even just a bit clumsy. Increasingly, it is acknowledged that the two most famous examples of people who stutter in the classical world are false. Demosthenes, the famous Greek orator, who overcame his stutter by putting pebbles in his mouth, probably had little more than a slightly weak voice and a lisp. Claudius, as we have seen, probably had Little's disease.

What is significant is that in each of these instances the individual in question is somebody attempting to participate in public life. Since this revolved around the traditions of rhetoric and logos, anything that impeded their flow, like a speech disorder, could – and did – prove a barrier to success. There was no question of educated society tolerating such a defect: not even Claudius, as Emperor, was free from relentless disgust and mockery about the disorder in his speech. The general view was that he was unfit for such service, but must be tolerated because of who he was. Only those who completely overcame a speech disorder, like Demosthenes, would be truly accepted. The vagueness of Greek and Latin terms is symptomatic of these attitudes: there was little point in closely analysing the speech of somebody unfit for public life because they were, after all, little more than barbarians.

Bizarrely, it seems that for the rest of the population, the vast majority who were not of aristocratic birth and unable to participate meaningfully in public life, a speech disorder was scarcely a problem at all. While it must be assumed that as many people had speech disorders as they do today, it is impossible to tell for certain because, as Laes says, 'difficulty in speech was not often used as an identifying characteristic'. In other words, it was ignored. This suggests it is only in the context of an elite culture which has come to revere fluency of speech that disorders first become problematic.

Over the following millennia, those societies across the world that have viewed themselves as inheritors of classical civilisation have automatically assumed a fluency-prejudice that is culturally determined rather than innate to human behaviour. In the eighteenth century, this prejudice became even more enhanced in Europe and North America with the perceived gulf between fluency and defective speech widening ever further.

There are many reasons for this shift, but the most important is economic. For thousands of years, European society had been predominantly feudal in nature with individuals born and dying in the same status. Social mobility, as we understand it, was extremely difficult. But from the 1500s, the merchant class we associate with Tudor England or the Florence of the Medicis had grown both larger and more powerful, and by the early eighteenth century, our economies depended on the fluid movement of capital as much as fixed ownership of property.

In the world of capitalism, historic class and feudal structures were starting to break apart: a person born into service might, with ambition and some luck, rise to the top and watch their former masters passing on their way down. The centres of this new economic order were the cities and the colonies. The increasingly large class of people born neither into land ownership or servitude traversed this landscape searching for opportunity.

The novelist Daniel Defoe's fictional heroine Moll Flanders charts this transformation: one that Defoe had witnessed over the course of his own life.[9] Born into penury, Moll becomes a prostitute in London, then a businesswoman and finally plantation owner in America: her status and wealth continually shift. At all times, within this fluid social order, she depends upon the gift of the gab as much as her sexual prowess. Fluent and charismatic speech, Defoe suggests, is what enables one to lie, impress, cut deals and survive.

For the emerging 'middling classes', who considered themselves above the hustle of the streets, but were also painfully aware of their inferiority to the aristocracy, mastering a certain register of speech was an important way of showing their worth. Books like Thomas Sheridan's *Lectures on Elocution* (1762) and John Walker's *Elements of Elocution* (1781) proved immensely successful. Elocution lessons were invaluable for businessmen of humble origin trying to navigate rarified circles, but also for women needing to pass in polite society or marry above their status. Unlike the schools of rhetoric of the classical world, which had included elocution as just one element in a more holistic approach to persuasion and political consciousness, these schools were simply about speaking well. While somebody with a speech disorder like stuttering can still prove competent at rhetoric, they cannot but fail at elocution unless they eradicate their disorder.

In 1750, Lord Chesterfield wrote to his eighteen-year-old son expressing immense concern at the 'hitch or hobble in your enunciation' which made him almost unintelligible. 'No man can make a fortune or a figure in this country without speaking, and speaking well, in public,' he writes. 'Your trade is to speak well, both in public and in private ... Be your productions ever so good, they will be of no use, if you stifle and strangle them in their birth.'[10] With the stakes so high, it is no coincidence that the first speech therapists, or 'artists', emerged around the same time.

In a recent study, Cambridge historian Elizabeth

Foyster dated the earliest practitioner to 1703, several decades earlier than previously thought. An advertisement states that James Ford, 'who removes stammering, and other impediments in speech', can be found every Tuesday and Thursday at Mr Merriden's, the sword cutler, during the day and at Rainbow Coffee House at six in the evening.[11] 'Speaking well was crucial to being accepted in polite society and to succeeding in a profession,' Foyster says. 'Speech impediments posed a major obstacle and the stress this caused often made a sufferer's speech even worse.'

Economic and social change explains the increasing emphasis on fluency in the modern era. And it is this aspirational speech, instrumental for social mobility, that joins with the deeper traditions of rhetoric and philosophy in creating what we recognise as hyper-fluency: the need to speak well at all times. Dale Carnegie, the American pioneer of self-improvement, saw this change clearly. 'In the days when pianos and bathrooms were luxuries,' he wrote in 1913, 'men regarded the ability in speaking as a peculiar gift, needed only by the lawyer, clergyman, or statesmen. Today we have come to realise that it is the indispensable weapon of those who would forge ahead in the keen world of business.'[12]

The new economic order, and the implications it had for speech, changed everything. Not only was fluency deemed a necessary trait for business, but the term became synonymous with the effective workings of capitalism itself. 'Flow', which shares the same etymological source as 'fluency' and is a common metaphor for admired speech, was requisitioned by economists trying to describe how capital works. We talk of 'cash flow', the 'flow of wealth', and 'stock and flow'. There are even asset management companies called Fluent Investments and Fluent Financial, while Deloitte Fluent Capital Adequacy is a piece of software that helps banks navigate financial regulations. In this way, what were once neutral, descriptive terms become virtues and we forget that all things that flow, like the course of a river, also require control and moderation. Dysfluency, by implication, like any impeder of flow, is not only antithetical to an individual's professional and personal fulfilment, but of little use in an increasingly fluid and globalised economy.

Dale Carnegie may have identified the shift in hyperfluency from a 'peculiar gift' to an 'indispensable weapon' as early as 1913, but, over a hundred years on, this change has only intensified. The collapse of the jobs-for-life ethos in the last fifty years, matched by the rise of the gig economy, means there is no exemption for anyone. With few of us able to know what we'll be doing a couple of years hence, we have to be prepared to impress at any given moment in interviews, meetings and presentations – and that means being able to talk well.

Joshua St Pierre, a Canadian speech activist and philosopher with a long personal experience of stuttering, describes to me the way twenty-first-century capitalism 'incites our tongue to speech the whole time'. It has transformed 'our capacities of speech into a form of human capital'. In so doing, it has compromised the effectiveness of speech as a tool of protest or change. While 'speaking up' was once daring to say the unsayable, whatever the consequences, now it is just part of capitalism itself: the very thing such protest is often directed at. Social media – the benign, collective term we give to vast global corporations like Google, Amazon and Facebook – depend upon us 'speaking up at all times and as much as possible' in order to fuel their data economies.

As Chris Anderson says, charismatic speaking is widely perceived to be 'a core skill for the twenty-first century'. How else do we explain the global success of TED, or indeed the relentless eloquence of social media, in which teenagers can become multi-millionaires delivering low-fi make-up or gaming tutorials from their bedrooms? What makes our twenty-first-century society exceptional isn't the existence of hyper-fluency, therefore, but its universal imperative: a core-skill not for a small priest class, but for us all. And because this society is increasingly globalised, it is ever harder to sit it out. In doing so, we have lost a great deal.

'Non-verbal communication is the highest form of communication,' Marina Abramović tells me, speaking English with her pronounced Serbian accent:

> The Buddhist teachers say that the most ordinary communication between master and a student is the former talking to the latter. Verbal communication. Then the second communication, which is much higher, is gesture. But the highest communication of all is sitting in silence with no words ever exchanged. So when I'm looking a total stranger in the eyes and not having any conversation with them, I will know more about that stranger than I will ever know through any conversation at length. There is something about

non-verbal communication that opens doors no other communication can open.

How do we get back what we have lost? How do we start to challenge our cultural reverence for a singular way of speaking; one which comes at the expense of tolerance for other ways of communicating and is, in any case, prone to misuse and error? Doing so involves sharing an awareness of the limitations of hyper-fluency, but also celebrating the virtues of different types of speech, even though hyper-fluency encourages us to be intolerant of them.

Later, I will describe the role speech disorders can have in challenging the cultural dominancy of hyper-fluency, but there is another story to tell first that explains the stigma that still surrounds them. It is the story of how, from the mid-nineteenth century, when the emphasis on fluent speech as the key to human happiness and success became most entrenched, disorders of speech in contrast became diagnosed, medicalised, and even treated with contempt.

5

A Muted History

Everything we think we know about speech disorders dates back a mere 150 years. Up until the mid-nineteenth century such conditions were either ignored, misdiagnosed or misunderstood. This doesn't mean they didn't exist: there are Ancient Egyptian texts from as early as 3000 BC connecting speech and communication problems to head injury,[1] and their existence in the classical world and beyond is (thinly) documented. While catch-all terms like the latin *balbus* gradually gave way to specific terms for stuttering or lisping, other disorders like dysarthria, aphasia and the vocal tics of Tourette's remained unnamed and were seen simply as symptoms of other problems: stroke or infirmity or some obscure palsy. But in the mid-nineteenth century, the increasing specialisation of medicine and advancements in our understanding of the human brain brought them into view. Of particular importance were the emerging fields of neurology and psychology which studied the brain from different angles: the first looking at the nervous system, the second at human thought and behaviour. Practitioners were asking questions that had never been asked before about how speech is produced and how it breaks down, and gathering evidence to answer them.

These investigations ran throughout the century, but in 1861 two important events occurred that proved hugely influential on our understanding of speech disorders today. They occurred in the cities of Paris and London: two great centres of science and innovation. One was a discovery made with a scalpel knife; the other a set of illuminating ideas contained in a book. These revelations, and those they inspired, were game changers, but they were frequently contested.

The first took place in Paris. Aged only thirty-six, Paul Broca was already a leading light of French medical science in 1861. Although a surgeon by trade, he had become the world's leading authority on aneurysms; made outstanding contributions to our understanding of cancer and the nervous system; and founded the Anthropological Society of Paris to extend the work of Charles Darwin in understanding human evolution. Along with his anatomist friends, Broca had become interested in human language and how it is produced. Was it something with an almost mystical origin that simply emanated from our consciousness, or could it be localised to a part of our anatomy? Those of a religious persuasion favoured (or hoped for) the former: the fact that no one knew where the faculty of language lay, left a space open to argue for the existence of the human soul. Those of more godless, Darwinian leanings hoped to dash their faith – if only they could uncover a bit of human flesh that seemed to be doing the job.

The search for the origin of human language, therefore, was more than an anatomical enquiry but an agitator on the fault-line between science and religion. As an ambitious and curious scientist, Broca hoped to claim this particular discovery for himself, sitting perfectly as it did across his skills as both an anatomist and anthropologist. The problem was how best to go about it. It wasn't acceptable to open the skull and dig around the brains of perfectly healthy individuals with a scalpel, hoping in the process to deactivate, and therefore locate, a language switch. Nor was it much use if those experimented on were already dead, as even if one did hit upon the right place there was no way of knowing since they were no longer able to speak.

Early in 1861, Broca was alerted to the case of a patient at Bicêtre Hospital in Paris who had an unusual condition. The man was a fifty-one-year-old farmer who had slowly lost control over his ability to speak, as well as experiencing paralysis through the right side of his body. His name was Louis Victor Leborgne, but he was known as 'Tan', because for many years it was the only sound he could make. Ask him his name, how he was doing, what he wanted to eat, and the answer was always the same: 'tan', 'tan' and 'tan'. In every other regard, his intelligence was unimpaired. For Broca, this strangely afflicted farmer presented a unique opportunity. Because Leborgne had already lost the use of his speech, it opened the tantalising prospect that if a fault could be found somewhere in his body, some damage causing the speechlessness, then the source of language itself would be located.

Broca had Leborgne, now bedridden and speechless, transferred to his care. He subjected him to tests to try and locate the source of his infirmity, but none was obvious. His nervous system was clearly working, even on the right side of his body, because of the 'flinching and screaming' that he made when Broca tested his scalpel on him. 'The tongue was perfectly free,' wrote Broca. 'The muscles of the larynx did not seem impaired at all, the timbre of the voice was natural, and the sounds the patient made in pronouncing his monosyllable were perfectly pure.'[2] Broca was at a loss. Then, shortly after these early examinations, Leborgne died. Within hours, Broca conducted an autopsy to try to understand once and for all where the root of his condition lay.

As Broca worked his way through the cadaver, he found Leborgne was in every way a perfectly functioning individual. Then he discovered deep lesions in the frontal lobe of the left hemisphere of his brain. This, Broca concluded, was the source. 'All evidence,' he said, in an address to the Anatomical Society of Paris later that year, 'leads us to believe in this case that the lesion to the frontal lobe was the cause of the loss of speech.' The implication appeared obvious: if a lesion to the frontal lobe could deprive an individual of speech, the same area must also be the origin of it.

In the following months and years, autopsies on patients with similar conditions seemed to confirm his proposal. This part of the brain soon became known, as it is called to this day, the Broca's Area. Speech deficiencies arising from damage to this area were called 'aphasia'. Broca had claimed both language and its disorders for neurology. And if aphasia could be diagnosed and explained, it was possible other disorders of speech might be too.

Across the Channel a different breakthrough, but of equal import, was under way. This was not a surgical innovation, but a way of seeing, a paradigm shift. The instrument was a book rather than a scalpel and it concerned stuttering rather than aphasia. The author was James Hunt, an extraordinarily precocious, prim-looking twenty-eight-year-old who had become the leading expert in stuttering. He was not a surgeon or neurologist; scarcely a scientist at all, although he had managed to scrape through a doctorate abroad to claim a qualification. He was one of the speech 'artists' who had emerged in London over the previous century.

While the origins of this profession undoubtedly had an element of hucksterism, with unqualified quacks reaping the benefits of the placebo effect from their supposed cures, it had gradually developed a degree of peer review among its members, becoming highly sophisticated in the process. Speech artistry continued to sit outside, often at odds, with the medical profession, yet by the mid-nineteenth century it was clear these practitioners were consistently having better results in the development of partial cures. Although we now talk of speech therapists, there was as yet no universal term for this sort of practice and little in the way of literature presenting a consistent method. Hunt was on occasion referred to as a 'psellis-molligist', although this ungainly term did not last long (probably for no better reason than the fact it is almost unpronounceable, particularly to those who might have recourse to use it).

James Hunt's *Stammering and Stuttering, Their Nature and Treatment* (1861)[3] not only summarised the collective achievements of this unofficial, almost underground profession, providing a blueprint for later speech therapy, but convincingly challenged the dominant conceptions of what stuttering is. Hunt was the first to chart the history of stuttering and the different ways it had been treated. In doing so, he showed how, as with other speech conditions, it had been repeatedly misunderstood. The first part of his book is a powerful chronology of human folly.

The earliest descriptions of stuttering, Hunt reveals, go back to the Old Testament, where Moses claims he is not fit to lead the Israelites because he is 'slow of speech and tongue', which has often been interpreted as a sign of stuttering. In the fourth century BC, the Greek physician and 'father of medicine' Hippocrates speculated that it is a consequence of disease and is accompanied by enduring diarrhoea. Hippocrates concluded that stutterers who are tall and baldheaded are good people; those with large heads and small eyes are passionate; and anyone with a little head will not stutter or go bald unless they have blue eyes. Celsus, another Greek philosopher, suggested the stutterer should exercise himself to retain his breath, wash the head with cold water, eat horse-radish, and then vomit. His contemporary Galen stated that stuttering is caused by either excessive moistness or dryness of the tongue.

The thread goes slack with the Dark Ages, but Hunt picks it up several centuries later, showing how little the understanding of stuttering had changed in that time. Guy de Chauliac, a medieval French physician, proposed blistering and bleeding as a cure. The Italian Girolamo Mercuriale agreed with the blistering, but also insisted that people who stutter should never wash their hair because it just adds to the excessive moisture in the tongue that is at the root of the problem. In the eighteenth century, Gottfreid Hahn identified the hyoid bone in the neck as the culprit; Anthony de Haen favoured a cavity in the lung; while Santorio Santorio blames the width of the incisive canal connecting the nose and mouth. Depending on who is writing, cures involve speaking with a bullet, a role of linen, a silver fork, a bride-langue (a sort of human bridle) or a whalebone in the mouth.

Again and again, Hunt identifies the misdiagnoses and inappropriate treatment of the medical profession. And for anyone convinced these are the errors of the distant past, he points to the recent atrocities of the surgical profession. Only twenty years before, Johann Friedrich Dieffenbach recounted the apparent success of his surgical cure on a group of teenage boys:

> The tongue being drawn as much forward as possible, I pushed a curved bistoury [scalpel] through it, as near its root as I could, and cut through its whole muscular thickness, leaving the mucous membrane inviolate … The substance of the organ was so completely cut through that a slight additional pull with the forceps would probably have torn it off. The blood streamed from the apertures made by the knife as vehemently as from a large artery.[4]

The results of these operations were, in Dieffenbach's words, 'beautiful'. He claimed his patients had not stuttered since. This apparent success encouraged other scalpel-happy surgeons across the continent until it became clear that any temporary alleviation of stuttering was simply a combination of placebo effect and shock that soon wore off.

The fundamental error, Hunt concludes, from Hippocrates to Dieffenbach, is a conviction that stuttering is physiological in origin, with the poor tongue more often than not getting the blame. 'There is perhaps no affliction to which the human frame is liable,' he writes, 'which has been attempted to be cured in so many different ways. The famous pebbles of Demosthenes; a bullet in the mouth; a roll of linen under the tongue; the fork of Itard; the bride-langue of Colombat; the whalebone of Malebouche; the stick behind the back; intoning; speaking through the nose; talking with the teeth closed.' Yet no physician had got any closer to solving the enigma.

Hunt's message, delivered within the first chapters of the book is clear: traditional medicine, having had ample time to find a cure for stuttering, had failed. It failed because it assumed there must be a single physical cause of the problem, usually the tongue, and because it took so little account of a patient's own experience and insights. But over the previous hundred years, according to Hunt, a different approach had slowly emerged alongside. It was developed by a combination of people who stuttered themselves, and fluent speakers who were willing to listen carefully to those who did. It was a way of thinking that would ultimately evolve into the practice we today call speech therapy.

The 'speech artists' of the eighteenth century tended to focus on an individual's habits of speech rather than their tongue, encouraging them to improve articulation and pronunciation[5]: simple techniques which help far more than whalebones, silver forks and surgery. And, increasingly, there were other individuals who believed the cause of stuttering had little to do with the speaking apparatus at all. The German philosopher Moses Mendelssohn suggested as early as 1783 that the main cause might not be physiological, but rooted in the experience and attitude of an individual. The English philosopher Erasmus Darwin (grandfather of Charles) concluded much the same. 'Impediment of speech,' he wrote, 'is owing to the association of the motions of the organs of speech being interrupted or dissevered by ill-employed sensations or sensitive motions, as by awe, bashfulness, ambition of shining, or fear of not succeeding.'[6] The cure involved practising 'for weeks or months upon every word, which the stammerer hesitates in pronouncing. To this should be added much commerce with mankind, in order to acquire a carelessness about the opinions of others.'

Significantly, both Mendelssohn and Darwin were also people who stuttered (as was Darwin's more famous grandson, Charles). Their theories came from personal insight rather than wild speculation. Because they foregrounded psychology over physiology, James Hunt considered them forerunners of his own way of thinking.

In the second part of *Stammering and Stuttering*, Hunt presents his own theory of stuttering. In his view, there is no single cause. It is not an organic disease at all, but a habit – and a bad one – of articulation, reinforced by the negative associations attached to it. In this regard, stuttering is primarily psychological, although the term Hunt uses is 'psychical'. He takes great pains to describe the mental condition of the stutterer: 'the habit of secrecy, of feeling himself cut off from his kindred; of brooding over his thoughts, of fancying himself under a mysterious curse'. All of which was reinforced by a medical profession that was not only completely confused in its diagnosis, wilfully spreading misinformation, but also infantilised people who stuttered by prescribing ill-conceived cures rather than listening to them.

Just as he rejects a single cause of stuttering, Hunt also rejects the idea of a single cure. His method consists of listening and studying the patient to understand how the habit had become acquired and what methods might be necessary to help them control it. His aim is to help an individual to self-awareness, clearing away the clutter of ignorance and misconception that surrounded the condition, building their confidence and helping them 'to speak consciously as other men speak unconsciously'. By learning about themselves and the mechanics of their speech, Hunt's 'pupils' (he preferred this word to 'patient') can steer their own path to fluency.

Putting theory to practice, James Hunt opened a country house residency near Hastings where pupils would come and study for weeks at a time. Each day, they would undertake certain activities together: reading aloud, debating and delivering speeches. In this safe environment, they could address the often insurmountable fear of public speaking, or even just everyday conversation, and build confidence and self-reliance as talkers. We all have a mental image, based on personal experience or movies, of what group therapy looks like – and it doesn't include men and women in Victorian dress.

Stammering and Stuttering did a number of things. It put to bed once and for all the idea that stuttering was a physiological problem that could be treated through surgery or any other tricks of the physician's bag. And it also comprehensively made the case for a multifactorial understanding of stuttering, recognising that the symptoms are psychological as much as behavioural. By focusing on literary as well as medical history (and the book is full of references to Shakespeare and the philosopher John Locke as well as the likes of Hippocrates), Hunt opened up the third dimension of stuttering, recognising the condition is shaped by the social and cultural attitudes of the time.

Finally, it was immensely influential in legitimising the emerging field of speech therapy. Hunt's pupils included some of the most prestigious writers and scientists of the day who contributed to the future integration of speech therapy into medical practice. Lewis Carroll and Charles Kingsley, author of *Alice in Wonderland* and *The Water-Babies* respectively, studied under him and were passionate believers in his methods. W.H.R. Rivers, one of the pioneers of British psychology, was Hunt's nephew and grew up in and around the practice. Without Hunt the legitimisation of speech therapy might have taken far longer than it did.

Hunt's thinking was not without its flaws. In ensuing years, he concentrated increasingly on the emerging field of anthropology. In 1863, he set up the Anthropological Society of London, where he delivered a disgraceful paper *On the Negro's Place in Nature* in which he defended slavery in the Confederate States of America. But in the field of speech therapy, which was his expertise, his impact and influence was almost wholly positive. *Stammering and Stuttering* remains a book of immense wisdom: the oldest practical guide to alleviating the symptoms of stuttering that continues to have merit.

In the same year, Broca and Hunt had made the case for both the neurological and the psychological roots of aphasia and stuttering respectively. These approaches are not mutually exclusive but complementary; today, it is widely thought that speech disorders are neurological in origin but subject to psychological factors. The three decades following 1861 proved a golden age with more achieved than in all human history up to that point.

An important successor in this endeavour is the towering figure of Jean-Martin Charcot. In the 1880s, he opened the first neurology clinic at Salpêtrière Hospital in Paris, gathering a group of brilliant physicians around him. Like Hunt, Charcot has a chequered legacy: his ill-conceived theories on female 'hysteria' created a toxic stereotype that lingers to this day despite being refuted within his own lifetime.[7] Charcot's real achievements lie in his work around neurological disorders. He was the first to define multiple sclerosis and Parkinson's disease, recognising dysarthria as an important symptom of both conditions. Under his guidance, students and colleagues made further breakthroughs. It was Charcot who named a newly identified 'malady of tics' after the groundbreaking work of his pupil Gilles de la Tourette.

As with multiple sclerosis and aphasia, there was no previous recognition that ticcing could constitute a disease or disorder in its own right. Jean Marc Itard had first described the symptoms in 1825, but the most wide-spread theory was that such conditions were exclusively an after-effect of rheumatic fever. This was not as ridiculous as it sounds: the tics of Sydenham's Chorea, known colloquially as St Vitus's Dance, were caused by fever. But at Salpêtrière, the young Gilles de la Tourette began collecting case studies of individuals who had similar types of tic but no experience of rheumatic fever. He became convinced of three things: that there was clear evidence for a malady of tics in its own right, that it was hereditary, and that it couldn't be cured.[8]

De la Tourette's theory struggled to catch on. Among his critics were the traditionalists who continued to insist such tics were an after-effect of rheumatic fever. Then there were those who said that 'la maladie de tics' was neither a syndrome nor an after-effect of Sydenham's chorea, but simply an elaborate manifestation of hysteria. Such florid hysteria was in itself a sign of advanced genetic 'degeneracy': a greatly discredited theory that argued negative mental and moral traits could be passed down from generation to generation, particularly among the poorer populations, leading to a weakening of the human species.

Gilles de la Tourette struggled to defend himself. He was severely debilitated after a mentally unstable patient shot him in the head, and suffered increasingly from mental confusion owing to syphilis. In 1899, he managed to publish an article salvaging what was left of his theory by conceding that the syndrome 'almost always exposed a condition of mental instability characterised by numerous phobias ... and all the stigmata which today are referred to as mental degeneration', although he preferred the term 'unbalanced'. What he would not endorse, though, was the notion that it was just an extreme form of hysteria. It was, always had been and always would be, a syndrome in its own right.

But the most damaging opposition to his theory ultimately came neither from the traditionalists nor the proponents of degeneracy, but from a new and emerging field that was to entirely change the way we understand the human condition. By the 1890s, some physicians and neurologists at Salpêtrière were fascinated by, and actively contributing to, the emerging discipline of psychoanalysis. They were close to Sigmund Freud who had briefly studied with them under Charcot in 1885. Freud, a neurologist by trade, was in turn influenced by the work they were doing into speech and other neurological disorders. It was while studying children with aphasia that he became convinced there were neurological and psychiatric conditions without any organic cause, driven instead by unconscious mental processes: a conviction that would become the bedrock of psychoanalysis.

One by one, the neurological breakthroughs in speech disorders that suggested they were linked to organic dysfunction in the brain were overcome by psychoanalytical theories. In 1891, Freud published his first book, *On Aphasia,* in which he dismissed Broca's claims of a localised disorder linked to brain damage as 'overrated' (he was right, but not for the reasons he supposed), leaving the field open for other theories, particularly his own. Over the course of the decade, he became increasingly emboldened. In *The Psychopathology of Everyday Life* (1904), he delivered a psychogenic reading of all forms of speech disturbance, or 'speech-blunders', where words are confused or inappropriately used. All such blunders, he suggests, are motivated by an unconscious drive such as repression. The famous 'Freudian Slip', as it became known, where one word is substituted for another, reveals 'self-criticism, an internal contradiction against one's own utterance'. Freud claims these principles apply also 'to those speech disturbances which ... affect the rhythm and execution of the entire speech, as, for example, the stammering and stuttering of embarrassment. But here, as in the former cases, it is the inner conflict that is betrayed to us through the disturbance in speech.'[9]

By linking major speech disorders with everyday slips of the tongue, Freud marginalised the role of neurology, foregrounding instead the role of a repressed and conflicted unconscious. People with 'speech disturbances' should not just be considered physical and verbal oddities, but psychological ones too.[10] None of this would have mattered if Freud's work had been forgotten, as so many of his contemporaries' treatises were, but *The Psychopathology of Everyday Life* proved one of the most influential books of the coming century.

For a long time, Freud and his contemporaries were viewed warily by the medical profession, but then thrust into the mainstream when a new ailment of the mind swept through Europe on an epidemic scale. In the First World War, hundreds of thousands of soldiers on the front line developed what we now call post-traumatic stress disorder (PTSD). The official term at the time was 'shell shock' and reactionary physicians believed it was literally caused by the impact of artillery explosions on the nervous system. For the still emerging field of psycho-analysis, shell shock was all too clearly a form of hysteria caused by the unbearable conditions of trench warfare. Because speech disorders were one of the most common symptoms, Freud and his followers had what amounted to a smoking gun connecting shell shock to psychogenic rather than neurological origins.

In *Psycho-Analysis and the War Neuroses* (1921), Sigmund Freud introduced the work of two of his dis-ciples, including the Hungarian psychoanalyst Sándor Ferenczi. In his article, Ferenczi talks of shell shock as 'a museum of glaring hysterical symptoms' including 'all the varieties of tic ... stuttering and stammering, aphonia [mutism] and rhythmical screaming'. With a certain tri-umphalist glee, he insists, 'There could be no question of a mechanical influence, and the neurologists have like-wise been forced to recognise that something was missing in their calculations, and this something was again – the psyche.'[11]

Over the course of the war, some army hospitals had reluctantly integrated elements of psychoanalysis into their practice, for the simple reason that it seemed effective in helping soldiers return to the front line. This compar-ative success, in contrast to the conspicuous failings of the medical profession, seemed a public vindication of a new discipline that had been widely considered an oddity. Emboldened, Freud and Ferenczi grew ever more cavalier in their speculations. Their great error was to mistake the benefits of therapy, where a professional and the patient talk through negative feelings, as vindication of their elaborate mythology of the role of repressed sexual and erotic functions on the behaviour of individuals. In one letter, Freud suggests to his colleague that stuttering is a displacement upwards of a psychological conflict over excremental functions.[12] Ferenczi, evidently not one to worry about a word like 'might', ran with the idea. In *Thalassa: A Theory of Genitality* (1924), he observed that stuttering on consonantal sounds suggested 'sphincter action, with anal inhibition'.[13]

Others followed in quick succession. 'Psychoanalysis of stutterers reveals the anal-sadistic universe of wishes as the basis of the symptom,' wrote Otto Fenichel. 'The expulsion and retention of words means the expulsion and retention of faeces, and actually the retention of words, just as previously the retention of faeces, may be either a reassurance against possible loss of a pleasurable autoerotic activity. One may speak, in stuttering, of a displacement upward of the function of the anal sphincters.'[14] Another psychoanalyst, Isador Coriat, saw stuttering as an extension of the suckling that an infant does on its mother's nipple. It is a form of mother fixation that explains 'the unconscious or latent homosexuality so frequently encountered in stammerers'.[15] The anxiety that people who stutter feel about their speech isn't a fear of embarrassment, 'but is a protective mechanism to prevent complete betrayal of the primitive oral and anal-sadistic tendencies through speech'.

If stuttering supposedly evidenced an erotic retentiveness, the phenomenon of ticcing suggested the opposite. Needless to say, psychoanalysis did not recognise ticcing as a neurological condition but simply as another symptom of a fretful unconscious. In 1921, Ferenczi published his 'Psycho-analytical Observation on Tic', in which he announced tics were 'stereotyped equivalents of Onanism [masturbation]'.[16] Coprolalia was 'nothing else than the uttered expression of the same erotic emotion'. Ferenczi drew great significance from a theory 'that tics often increase in power at the time of early puberty, pregnancy and childbirth, at the time therefore of increased stimulation of the genital regions'. The presence of coprolalia simply confirms 'the impression that the significant "displacement from below upwards" so strongly emphasised in neurotics as well as in normal sex development plays no inessential part in the formation of tic'. Astonishingly, in reaching these conclusions, Ferenczi confessed he had never examined a patient with profound tics but drew instead on the reports of others.

Yet, as a form of treatment, psychoanalysis seemed to have little effect on ticcing. The more it failed, the more outlandish its claims were. In the 1940s, Margaret Mahler, an exile from Nazi Europe living in New York, published a series of psychoanalytical studies of children with 'tic syndrome'.[17] While she acknowledged a hereditary predisposition in particular individuals, it remained latent, she claimed, unless brought out by unique psychosomatic conditions. According to Mahler, all people with the syndrome were emotionally immature with a 'preoedipal mother fixation' and 'homosexual claims directed towards the father'. She described the ticcing of the various patients she encountered as deriving from a 'masturbatory conflict', or a 'psychosexual conflict on an anal-sadistic and masochistic level'. And yet, having done extensive first-hand research (unlike Ferenczi), Mahler was forced to conclude that there was 'no direct correlation between recovery from the tic syndrome and length or method of psycho-therapy'.[18]

In an astonishing twist of logic, Mahler claimed this was of no detriment to her profession: tics were a way of releasing sexual and unconscious tension. If the symptoms of the syndrome disappeared, then that same tension usually resolved into a 'severe personality maladjustment'. The role of psychoanalysis, therefore, was not to eliminate tics, because they might be the last bastion preventing a complete disintegration of an individual's personality, but to tackle the underlying masturbatory and psychosexual conflict that Mahler believed was causing them. In her case studies, she repeatedly engages young, often prepubescent, children in prolonged and explicit interrogations about sex in a manner which, from a twenty-first-century perspective, borders on abuse.

Retrospectively, it is easy to see that while none of these individuals intended harm, they were locked in a culture of groupthink – of conferences, periodicals and personal correspondence – that led them from one insupportable assumption to another. The astonishing thing isn't that they wrote such nonsense, but that these theories gained mainstream credibility. For most of the twentieth century, the Freudian or psychoanalytical understanding of stuttering and Tourette's syndrome was the dominant one in both the medical and the popular imagination. Even the treatment of aphasia, a condition so clearly linked to brain damage, was impacted. For a while after Freud, there was a tendency not only to address the impairments of speech in a patient but what one psychotherapist called their 'total personality',[19] as if aphasia offered the opportunity to perform some general housecleaning of the soul.

Under the influence of psychoanalysis, the popular perception of a person who stutters was, and remains, that of an individual with an excess of neurotic tendencies. There's the stuttering Billy in Ken Kesey's novel *One Flew Over the Cuckoo's Nest* and Miloš Forman's iconic film adaptation, whose speech impediment is a sign of his repressed Oedipus complex and harbinger of his eventual suicide. Nicholas Mosley, son of the British fascist leader Sir Oswald Mosley, spent many years in psychoanalysis trying to understand his speech impediment. 'A stammer is the indeed often ludicrous outward sign of an inward contradiction,' he wrote in his autobiography.[20] He also believed 'it is a protection against the stammerer's own potential aggressiveness towards others.' No wonder that sociologist Erving Goffman, writing at the height of psychoanalytical influence in the 1950s, listed stuttering as one of those 'unmeant gestures' like stumbling, belching and flatulence, that embarrassingly expose one's private self in a public arena: when you stutter, you're advertising your most intimate neuroses to the world.

Whatever the successes of psychoanalysis in treating other conditions, it remained the case that it just didn't work for speech disorders. Worse, it actually hindered our understanding of them, pushing to one side some of the advancements made by neurologists and speech therapists in the late nineteenth century. By the 1950s, an increasing disillusionment sent professionals scavenging for other cures. While James Hunt reserved his worst scorn for the surgeons of his own generation, one wonders what he would have made of some of the atrocities to come. Not only the excesses of psychoanalysis, but passing fads like the carbon dioxide therapy of Hungarian psychiatrist Ladislas Joseph von Meduna,[21] or a brief experiment in lobotomising children with Tourette's syndrome (in a later study, it was found that of sixteen such children, only five were deemed to have improved: the remaining eleven appeared worse than before the operation).[22]

Unsurprisingly, by the 1960s there was a growing resistance to psychoanalytical theories of speech disorders. This was partly due to the failure of empirical evidence to support them, reinforced by the fact that psychoanalysis itself offered little in the way of effective treatment. A new generation of neurologists and psychiatrists began to look elsewhere for solutions. Broca's theory about localisation, that both the faculty of speech and the cause of aphasia lay in the front lobe of the brain, had lost influence in the first half of the century, but was now reappraised. This was largely due to American neurologist Norman Geschwind who found that, despite its simplicities and flaws, there was more evidence in favour of what he called 'the classical neurologists' (Broca and Carl Wernicke) than the 'holistically oriented neurologists' of the twentieth century who integrated elements of psychology and psychoanalysis into their practice.[23] While nobody would argue that speech sat exclusively in Broca or Wernicke's areas, they were clearly important in speech production and damage to them, as well as other parts of the brain, provided a better focus for neurologists than the mysterious workings of the unconscious.

Stuttering was also slowly being reclaimed from psychoanalysis. The independent speech artists of the nineteenth century had always sat to one side of the medical profession, if not in outright hostility. After the First World War, speech-language therapy, as it was now called, was increasingly integrated into hospitals and universities. Not least because the number of returning soldiers with significant speech disorders, whether caused by shell shock or brain damage, required it.

Wendell Johnson and Charles Van Riper met in the 1930s in the newly-formed speech pathology department at the University of Iowa. Both men had stutters themselves and so were instinctively wary of some of the excesses of psychoanalytical theory. Instead, their ideas show the influence of Behaviourism, a psychological method which focuses on an individual's environment and past experience. The important thing wasn't to waste hours dredging the unconscious for a cause of stuttering, but to create a more constructive environment and support network for the stutterer, and to give them techniques for improving fluency. They developed a range of practical techniques to support this. These included voluntary stuttering which desensitises the experience of stuttering and builds confidence by making an individual stutter on purpose, but in a controlled way. Another was stuttering modification therapy which enables a person to tackle difficult words through a softening and slowing of sounds. These techniques were simple but effective and had a far greater impact than psychoanalysis ever achieved.

But the greatest setback for psychoanalysis followed the discoveries made about dopamine in the 1960s.

Dopamine is a natural chemical that works as a neuro-transmitter, sending signals throughout the brain and the body. Nobody knew a great deal about it, but it was found that drugs which reduced or enhanced dopamine impacted a wide range of neurological conditions. In his book, *Awakenings*, Oliver Sacks described the experiments he was involved in at Beth Abrahams Hospital in the Bronx. Sacks and his colleagues gave the dopamine precursor, L-DOPA to patients who had been immobile with *Encephalitis lethargica*, or 'sleeping sickness', for decades. Astonishingly, they emerged from their catatonic state, but soon developed violent side effects, including ticcing. The connection between neurotransmitters and tics caused pause for thought: had Gilles de la Tourette been right after all? Was it possible that 'la maladie de tics' wasn't psychosomatic, but a neurological condition?

Around the same time, two married doctors, Elaine and Arthur Shapiro, began prescribing haloperidol, a drug that reduces dopamine transmission in the brain, with startlingly successful results. In 1968, they published their findings, directly challenging the psychoanalytical community, arguing that the success of haloperidol suggested ticcing was not psychogenic in origin.[24] Gilles de la Tourette, they concluded, had been on the right track in the 1890s before psychoanalysis undermined his findings. By giving the illness a name – Tourette's syndrome – and insisting it was a neurological condition that could be pharmaceutically treated, parents of children with tics found themselves overwhelmed with relief. One father wrote of his joy 'to finally meet with an unusually competent psychiatrist who stated with assurance that the tic symptoms did not indicate that our son was psychotic, neurotic, or emotionally disturbed because of his family environment and parental inadequacy'.[25] The Shapiros joined with a handful of such parents to form the Tourette Syndrome Association which over the 1970s led a sustained media campaign, placing adverts and articles that proclaimed these findings and building a vast support network across the United States.

The importance of dopamine quickly emerged in other disorders affecting speech. Haloperidol has been effective in the treatment of stuttering, although it is rarely prescribed because of the side effects. The Austrian biochemist Oleh Hornykiewicz became convinced of the link between Parkinson's disease and dopamine deficiency as early as 1961. When he first experimented with L-DOPA on his patients, he was astonished to see them stand up and walk. 'Speech became better,' he recalled, 'they started laughing and actually crying with joy.'[26] It remains one of the few effective treatments available today, although the side effects, ranging from involuntary writhing to impulse-control disorder, can prove as unpleasant as the disease itself.

Whether prescribed or not, the effectiveness of dopamine enhancers or inhibitors was vital in diminishing the psychoanalytical hold over speech disorders and reclaiming them as neurological conditions. In effect, the 1970s saw a resumption of the work that had begun in the golden age of neurology, lasting thirty years from Broca's and Hunt's breakthroughs of 1861 to the emergence of psychoanalysis in the 1890s. This meant that, apart from the overdue recognition and integration of speech-language pathology as a practical discipline, decades of progress had been compromised.

Even today, the Freudian paradigm doggedly refuses to quite lie down. Every time we talk about someone having 'a nervous tic', we are reinforcing a notion that tics are caused by anxiety. 'They are,' says Jess Thom, 'but they're also more pronounced if I'm excited or happy. *Any* heightened emotion has an impact.' Before I was married, I remember going on a date with a woman who, after hearing me stutter repeatedly, squinted her eyes and asked, 'Are you a very angry person?', as if my impediment was nothing more than a sign of some repressed and unconscious fury.

Many people who stutter unwittingly contribute to trauma-based theory, not realising that in doing so they encourage people to think of stuttering as evidence of some emotional or neurotic crisis. Most people I speak to have their theory: poet and rapper Scroobius Pip tells me his speech disorder arose after nearly drowning as a boy; author Colm Tóibín says that his appeared when his father almost died; artist Brian Catling believes it may be the result of a teacher forcing him to be right-handed; and the writer Margaret Drabble was told by her mother that her stutter came on after she fell in a river. But while therapist Uri Schneider has encountered a couple of people who have undeniably trauma-induced stutters, he tells me it tends to be 'trauma with a capital T' like battle fatigue. Otherwise he's not so sure. 'Sometimes the story is that there was a dog that barked and mummy wasn't holding Johnny's hand,' he says. 'We want to make sense of things.

There's a human desire to understand and come up with a story that seems plausible because the worst thing is to sit with something as an unanswered question. People are looking for answers.'

We still are. The history of speech disorders is a long and painful narrative of misguided diagnosis and treatments that have tended to do more harm than good, failing to significantly alleviate the physiological symptoms while contributing to a stigmatisation which is, after all, the biggest problem for most sufferers. For me, the insights of neurology and psychology first developed in the late nineteenth century, then largely ignored, and finally integrated over the last three decades, are the most significant periods of advancement in our knowledge of speech disorders.

There are clear lessons to be drawn from this obscure history. The first is that we should never doubt the role of cultural determinism in the labelling of speech disorders. The fact that most didn't even have a name until relatively recently, or that their very existence was contested, confirms it. The second is that many of the greatest insights and breakthroughs come from personal experience. These may be from people with conditions themselves, such as Moses Mendelssohn, Erasmus Darwin, Wendell Johnson and Charles Van Riper. Or they may be through the close partnerships between experts and those with conditions, as in the examples of James Hunt and his 'pupils', or the Shapiros and the parents of children with Tourette's.

A third lesson is that we should never under-estimate the accumulative effect of centuries of misdiagnosis. It casts a long shadow over the way we hear and respond to speech disorders. Changing public perception will require tackling the legacy of psychoanalysis and proving that speech disorders have nothing to do with repressed emotions and disintegrating personalities. Most of all, we should be extremely wary about accepting any contemporary orthodoxies about the causes and cures of speech disorders. The surgeons of the nineteenth century and the psychoanalysts of the following were just as certain about their scalpels and anal fixations, yet have been rightly discredited. What appears convincing today may be equally discredited in a hundred years time.

6

Unfinished Stories

The history of speech disorders may be a catalogue of misdiagnosis and ill-conceived treatment, and obviously so when considered with the benefit of hindsight, but that does not automatically make the present an age of enlightenment. Our understanding of these conditions continues to change by the year. There remain more unanswered questions than answers. And broader awareness in our society, which is the key to acceptance and rehabilitation, is next to non-existent. Even in my lifetime, theories and treatments have come and gone like seasonal fashions, making it hard even for people with speech disorders to make sense of their experiences. When I look back, I see myself drifting between different approaches, sometimes achieving a brief improvement in my fluency and confidence, then adrift once more.

Although I have memories of stuttering from a younger age, it only became a significant problem for me around the age of ten. The mid-1980s was, however, not a great time to be diagnosed with a speech disorder. Psychoanalytical assumptions about stuttering lingered on and it was still seen as a sign of inner conflict and weakness that could benefit from a firm hand. Implicitly, there was a sense that many people who stutter chose to do so, if only at an unconscious level, and that if they toughened up or sorted out their hang-ups it would no longer prove such a problem. I think this is why laughing at stuttering was actively encouraged within society. This was the time, after all, when the UK's favourite sitcom (*Open All Hours*), its Oscar-winning film (*A Fish Called Wanda*[1]) and one of its hit singles (*The Stutter Rap*) all hinged around characters with supposedly hilarious stutters performed by people without them.

'Interviewers turn away, who wants to be covered in spray?' – rapped Morris Minor and the Majors in *The Stutter Rap*. 'Talkin' to me for more than an hour is equivalent to an April shower.' The song reached number four in the UK singles chart. It was an unfortunate backdrop to my early experiences of stuttering. When I blocked on a word at school, the class would roar with laughter and impersonate me. I learned *The Stutter Rap* in self-defence. If somebody quoted a line, I could pick it up and finish the verse, then change the subject.

The story of my treatment is preserved in my mother's newspaper cuttings folder. 'By 10, he had become almost incomprehensible,' she stated, rather bluntly, in an article for the *Guardian*.[2] I was taken to see a local speech therapist, but have few memories except a sense of hopelessness. I remember hearing about Demosthenes and his wretched pebbles, a few tips on how to slow down, followed by silent drives home. 'Months of treatment seemed fruitless,' my mother wrote, and I was deemed problematic enough to be referred to an intensive two-week course at the Farringdon Health Centre.

What I didn't know when my parents and I arrived at a rather shabby pre-fab building in central London was that we were guinea-pigs in a radical new treatment that was being watched with fascination by speech-language therapists across the world. I was one of eight children with pronounced stutters, all accompanied by both parents.

The course at Farringdon was led by three women – Lena Rustin, Willie Botterill and Frances Cook – who brought different skills and experiences to bear. They had discovered what all therapists know: it's easy to make someone fluent in the consulting room, but almost impossible to make it stick outside. It's the everyday environment (family, school, the work place) that reinforces stuttering behaviour. The very format of therapy, the lone subject with the lone specialist, was therefore both a help and a hindrance. What they needed was a form of therapy that would not only change a stutterer's speech, but the day-to-day environment in which they moved. But that's not something that can be done by sitting down for an hour once a week.

'Lena made the case,' Willie Botterill tells me, when we meet up thirty years later. 'She was the pioneer. She managed the whole National Health Service in Camden and Islington. So she had a big organisation and could decide how the money was spent. And because stammering was her particular thing, she prioritised it.' The course they founded did three things. First, it brought children together in group therapy so they could develop a peer-to-peer support network in the following months. I was twelve years old and remember feeling immensely safe in those over-lit, barely furnished institutional rooms, but also part of something edgy and experimental. Second, it applied Fluency Shaping therapy, taught to us as Smooth Speech, through which we learned to speak from scratch. I recall Willie recording my speech, working out that I spoke well over 200 words a minute, and feeling very special when she told me that only John F. Kennedy spoke as fast.

Finally, and most significantly, it kept all the parents in another side of the building for the entire two weeks, where they were both scrutinised and taught to recognise and change negative environmental factors. 'Impatient with Jonathan's fumbling fast-forward narratives,' my mother wrote (this time for *Good Housekeeping*), 'I'd done what so many people do with stammerers, and looked away when he spoke; and he, in response, kept his eyes on the floor when talking. To help our children become fluent we had to examine ourselves and the way in which we communicated as a family. This meant becoming aware of basic social skills, listening, turn-taking, praising, problem-solving.'[3]

This was the spring of 1987. It was the first time a group of therapists had attempted both behavioural and environmental changes on such a scale in the UK. And the results were spectacular. None of us were 'cured', but we became more fluent. Most importantly, we found ourselves more resilient and with savvy, supportive families who were able to connect with how we were feeling, how we were being treated at school, and intervene appropriately. A few years later, an evaluation of one cohort showed that around 45 per cent of children became, to all extents and purposes, 'normal' speakers with fewer than three stammered words a minute.[4] The work of the Far-ringdon Health Centre (now the Michael Palin Centre for Stammering Children) was widely celebrated and repli-cated in different ways around the world.

I assumed I'd experienced the last word in speech therapy. As a young adult, I would practice 'smooth speech' whenever my stutter got particularly unruly. But in my early thirties, I felt overwhelmed once more and feared it was negatively impacting my work. I got in touch with Willie Botterill, having not seen her for over a decade. It was strange returning to the same building and rooms I had spent so much time in as a child. Willie rec-ommended I sign up for the adult speech therapy group at the City Lit Education College. After an assessment with Carolyn Cheasman, I was placed in the 'interiorised stammering' class.

In those first sessions at City Lit, a couple of things took me by surprise. The first is that stuttering was no longer talked about as a predominantly psychological dis-order, but a neurological one: something different in the wiring of our brains. We, a group of around ten adults, were bluntly told that there was no cure and, in any case, that wasn't the point of therapy. The important thing was to change our thinking so we didn't feel so negatively about it. In place of Fluency Shaping therapy, the 'smooth speech' of my childhood, was seemingly the reverse: vol-untary stuttering. We began by having conversations in pairs, during which we had to fake a stutter while main-taining eye contact. This latter part was essential: over the decades, we had learned to always look away when stut-tering as if caught in the performance of some terrible, shameful act. Eventually, we were sent outside in pairs, where we had to accost strangers, asking for the time or directions, and deliberately stutter. In doing so, we couldn't break eye contact or apologise for our speech. I found this horrifying at first, but it soon became easy and, as a result of worrying less, I eventually stuttered less.

What stuck with me most was how much the emphasis of speech therapy seemed to have changed since my childhood. The shift from smooth speech to voluntary stuttering, for instance, felt a complete turnaround. I didn't know whether this was simply a reflection of the difference between treating children and adults, or whether something more fundamental had changed within the discipline. When I ask Willie about this, she suggests it was a bit of both:

> When we started out, it was simple, our job was to get you fluent. Now it's different. How much you stammer is not particularly important. There are people for whom the physical act of being fluent needs so much management that it compromises other things and they end up not communicating in the way they'd like to. If you can get to the point where the amount of fluency is not the issue and communication becomes the important bit, then that substantially alters both the way the individual and the world around them manages and thinks about it.

This change, from a practice focused on curing to one of individual empowerment, is perhaps the biggest in the history of speech therapy. It is not an absolute reinvention

– Fluency Shaping therapy still remains popular, while voluntary stuttering dates back to the 1930s – but it is a shift in balance that is only increasing. One can speculate as to the different reasons for it, not least a general shift in many societies towards greater tolerance for difference or diversity. But I believe the most important factor is the growing acceptance that most speech disorders are neurological conditions antagonised by secondary, psychological considerations. This change is significant because it means that stuttering and ticcing are no longer signs of a disturbed unconscious, or embarrassing 'slips' in an individual's front (to recall the formula of Erving Goffman), but just something different in the structure and workings of the brain.

If most speech disorders are neurological conditions, then the whole notion of trying to conceal them through mitigating techniques becomes morally dubious. There may be people within society who still believe that those who are disabled or different should be invisible, but they, rather than those they despise, are what needs most to change. It explains too why the emphasis of speech therapy has moved to empowerment, changing the way a person thinks and feels about their disorder, rather than trying to suppress its existence. This shift is still recent history. Howard Kushner, author of *A Cursing Brain?*, recalls being convinced, as so many others were, that Tourette's syndrome was psychoanalytic in nature as late as the 1980s. The tipping point for both stuttering and Tourette's occurred through the following decade.

The change in emphasis has helped to reduce the stigma of some conditions, but there is a risk that, unless clearly defined, the word 'neurological' becomes as jargonistic as all those terms – from 'psychogenic' to 'narcissistically-fixated' – that psychoanalysis thrived upon. To take one definition, this one from the World Health Organisation: 'Neurological disorders are diseases of the central and peripheral nervous system. In other words, the brain, spinal cord, cranial nerves, peripheral nerves, nerve roots, autonomic nervous system, neuromuscular junction, and muscles.' Neurological conditions are compounded by psychological considerations. After all, anxiety, pessimism and despair can exacerbate the symptoms of almost any ailment through obsessively thinking about them. Yet such conditions are not in themselves caused by negative emotional states, but by fundamental differences in the brain and wider nervous system.

Since aphasia and dysarthria are usually secondary consequences of brain injury or progressive disorders like Parkinson's, their neurological rather than psychiatric origin seems beyond dispute – and, to an extent, has been since the late nineteenth century. Stuttering and ticcing, however, have no obvious cause. Studies since the 1960s have shown how changes to the levels of dopamine in the brain can both exaggerate and diminish ticcing and stuttering, just as it affected the dysarthria of neurological conditions like Parkinson's. Then, from the 1980s onwards, advances in neuroimaging technology made it possible to peer inside people's brains. Again and again, scans show that there are significant differences in brain structure and activity between 'fluent' speakers and those who stutter and have Tourette's syndrome.[5] Yet compared to other fields of medicine, such technology is still in its infancy and the role of neurology remains theoretical. 'There are still many inconsistencies across studies,' concludes one recent report, 'and a comprehensive understanding of the specific mechanisms underlying Tourette's syndrome remains incomplete.'[6]

At the Oxford Centre for Human Brain Activity, Dr Kate Watkins is leading a cutting-edge programme called INSTEP (a near acronym of Investigating Noninvasive Stimulation to Enhance Fluency in People who Stutter) looking at the brains of people who stutter. I went to spend an afternoon shadowing her team at work. In one room, a volunteer from Bristol called Ahir waved cheerfully at me before rolling back into a vast state-of-the-art scanner. In the observation room next door, I watched his brain light up as he performed a series of verbal exercises. 'What we're finding,' Watkins tells me, 'is that people who stutter have a difference in the organisation of the white matter: the fibres and fatty insulation connecting the different brain areas that are needed to communicate. So it could be that the fibres are less well connected in stutterers or not as well insulated. What's significant is where we're finding this difference: close, almost in, the Broca's area.' Although Broca's theory of a single location of language has given way to one which emphasises the role of modules throughout the brain, the Broca's area is still considered of vital importance to speech production.

Researching this neurological difference is notoriously problematic. While many scans have been made of the brains of people who stutter when at rest, it is extremely difficult to conduct them during the act itself. The scanner is noisy, rhythmic and isolating: stimuli that generally make people stutter less. From what data Watkins has managed to produce, there is a great deal of activity in both sides of the brain at the moment of stuttering, but what is prompting this firework display – whether it is the cause of the stutter or the brain's reaction to the experience – is uncertain. And the evidence of neuroimaging doesn't quite explain the extreme variability of the condition; why a person who stutters might be unable to speak one moment and entirely fluent the next.

Watkins suggests one way to think about it is as a tendency, like a weak knee: 'When you're tired or anxious, it gets worse and reinforces itself. If one day you struggle and can't say your name, the next time you have to say your name, you are going to get more anxious. It becomes a vicious circle. Anxiety is not the cause of stuttering, but the result of having stuttered.'

Even the mystery of how people who stutter can become fluent when speaking in a foreign language, or singing or acting, can ultimately fit within a neurological framework. 'When you are speaking in a different language or as someone else, like an actor,' Watkins says, 'you're switching from your habitual manner of speaking into another one. This engages another part of the brain in speaking: a part of the brain which might work quite fluently. And when you revert to your habitual part of your brain, you start stuttering again.'

Brain scans of people with Tourette's syndrome have enabled similar discoveries. According to Tourettes Action UK, 'some structures in the basal ganglia part of the brain and in the fronto-temporal brain areas' are different in individuals with TS. But while neuroimaging can show difference in the structures or connections of the brain, it can't explain why. In the case of people who stutter or have Tourette's, it seems that people are simply born with it. This is reinforced by genetic research that suggests both a stuttering and a ticcing gene, and by the anecdotal evidence that both run in families. 'Around half of our cases report a relative who also stutters,' Kate Watkins says. 'And there have been mutations identified in the GNPTAB gene which seem to correlate to stuttering.'

While these breakthroughs are all significant, there are those who caution against assuming too much. 'I think we're moving the ball forward but it's still highly enigmatic,' says American speech-language therapist Uri Schneider. After all, it isn't beyond doubt that the differences in brain structure and connections are the cause of a speech disorder rather than a consequence of them. The brain is highly adaptable and can change to match human behaviour as well as being a cause of it. The genetic work is, likewise, promising but limited. 'There are many stutterers who do not seem to have any genetic history,' says Schneider. 'And while many do, you can have genes for all kinds of things but that doesn't mean they're activated. Ten, fifteen, twenty years, we may be there in terms of understanding the true nature of stuttering, but we're nowhere near yet.' And, of course, none of the scans help explain one of the biggest mysteries of all: why there are up to four times as many men with stuttering or Tourette's as women.

Unsurprisingly, caveats and uncertainties continue to run through the literature about speech disorders. Many bold claims are saddled with conditionals: the devil is still in the small print. 'Tourette Syndrome (TS) is an inherited, neurological condition,' begins *What Makes Us Tic*, an explanatory leaflet provided by Tourettes Action. This is an important statement because it helps shift societal prejudice away from notions that sufferers are psychologically disturbed or deliberately provocative individuals. But this firm claim is soon moderated: 'Although the cause has not been established, it appears to involve an imbalance in the function of neurotransmitters (chemical messengers in the brain), dopamine and serotonin. It is also likely to involve abnormalities in other neurotransmitter systems of the brain.' As for the hereditary roots, it concludes that 'the genetic cause of TS is complex as not one single gene has been identified to be the cause of the condition.'

Official statements about stuttering tend to remain even more cautious. According to the NHS website: 'It isn't possible to say for sure why a particular child starts stammering, but it isn't caused by anything the parents have done. Developmental and inherited factors may play a part, along with small differences in how efficiently the speech areas of the brain are working.' *The Handbook of Language and Speech Disorders* acknowledges there is probably 'an underlying neurological cause', but emphasises that most theories today are 'multifactorial': stuttering is a perfect storm of genetics, brain structure, environmental factors and the unique psychological response of each individual.

To the physicians of antiquity through to the nineteenth century, multifactorial theories would have seemed wishy-washy: they wanted a single, smoking gun (or, in this case, tongue). While posterity may disfavour this current tendency, just as it has those of the past, the advantage of multifactorialism is that it is non-exclusive, permitting and accommodating new theories. To take one example, Naheem Bashir, an experimental psychologist at the University of London, is researching the extreme variations in stuttering behaviour within an individual's own experience. When I saw him speak at the Wellcome Collection in London, he described it, as it generally is, as 'neurological, with a genetic and developmental base', but his use of neuroimaging has uncovered an intriguing characteristic.

While brain activity in fluent speakers is similar in most speaking scenarios, Bashir claims it varies significantly for people who stutter depending on whether they are speaking alone or speaking in company. When speaking alone, their brain activity is the same as those who are fluent, but it transforms when they're speaking in company. It isn't the act of speech, therefore, that prompts unusual activity in the brain but the act of interaction. To explain this, Bashir suggests stuttering could be linked to what is called 'social information processing', or how our brains interpret and react to social situations. Whatever the neurological and genetic basis really is, stuttering appears to be activated by certain social scenarios. It may not have an on-off switch, but it seems to have a dial that goes from low to high. The same is true to an extent with Tourette's syndrome in which both the frequency and very nature of tics are influenced by social context.

Greg Snyder, a Professor of Communication Sciences and Disorders at the University of Mississippi, goes one step further. He highlights a small number of studies over the decades suggesting that speech disorders occur in sign as well as spoken languages. While it is easy enough to conceive how the articulation difficulties of dysarthria (linked as they are to wider motor disabilities), the cerebral disconnect of aphasia and the tics of Tourette's all translate into Sign, there is evidence that the involuntary repetitions, prolongations and blockages of stuttering do so as well, enacted by the hands and body rather than the mouth. Yet it is impossible to account for why this is the case, unless we are willing to acknowledge that stuttering is more than a speech disorder but a condition that affects communication skills more generally.

The enigma causes Snyder to argue that 'the traditional views and definitions of stuttering as a speech disorder fail to account for the stuttering phenomenon', for while they account for stuttering as it occurs in the mouth, they fail to accommodate or even acknowledge similar behaviour in the gestures of a Sign user.[7] His recommendation is that researchers and clinical scientists should consider 'abandoning much of the prevailing paradigmatic thought on stuttering ... a new paradigm will need to emerge to account for this new perceived reality.'

As theories about the cause of speech disorders have changed, so too have the recommended treatments. Looked at over time, speech therapy can appear remarkably faddist in nature. The methods of Lionel Logue in the 1940s seem odd but harmless, while those of psychoanalysis misguided and often cruel. Of far greater and enduring credibility are the methods that emerged in the American midwest from the 1930s. Stuttering modification therapy is associated with Charles Van Riper, but early variations of it were practised by nineteenth-century speech artists like James Hunt. The course I attended at what would eventually become the Michael Palin Centre for Stammering Children owes a debt to the theories of Wendell Johnson, the architect of the controversial Monster Study, who emphasised the importance of changing the environment and mindset of a person who stutters, rather than their speech alone.

The Lidcombe Programme in Australia, which started around the same time, boasts immense success in promoting fluency by teaching parents to congratulate and therefore reinforce fluent behaviour. And since the 1990s, the McGuire Programme in the UK has become extremely popular, combining breathing techniques and sports therapy with exposure therapy. Participants are given a series of increasingly daunting public speaking challenges, from cold-calling on the telephone to addressing a crowd of strangers in public spaces like Trafalgar Square or in whatever urban centre the course is being run. It instigates a highly emotional and dramatic journey of personal transformation, although critics suggest it doesn't address the core problem, and improvements in fluency are often temporary.

There are technological tools too. The SpeechEasy is an audio device that has helped many: rather than increasing volume like a hearing aid, it loops and adjusts the sound of your own voice creating a 'choral effect' as you speak. In another of the strange variables of stuttering, this is enough to render some speakers mostly fluent. Kate Watkins is achieving some promising results with transcranial direct-current stimulation, which involves running a very low electrical current through the brain. Pharmaceutical treatment is rarely endorsed because of the side effects, but there is evidence that dopamine reducers[8] and SSRIs (selective serotonin reuptake inhibitors) like paroxetine[9] are effective. But both technological and pharmaceutical treatments are increasingly out of favour, partly because they are cumbersome interventions but also because they fail to address the core of the problem, which is not the individual's speech but how their attitude and their environment exacerbate it.

While certain practitioners and programmes will talk, as they have always done, about providing a cure for stuttering – even presenting the testimonies of individuals to support this – the reality is that there isn't, and never has been, such a thing. If current theories about the cause tend to be multifactorial, the most effective treatment tends to be as well, rooted in pragmatism rather than ideology. 'There's no one size fits all in therapy,' Uri Schneider tells me. 'We're not surgeons, we're there more like a midwife to help someone through a challenging process with a lot of guidance, support and tips. But ultimately it's their journey and we're there to be along for the ride.' Schneider's method involves taking all the tried-and-tested techniques of speech-language therapy over the decades, without any ideological prioritisation depending on which school of thought they support, and offering them as an armoury to choose from.

Any hope of a miracle cure for the vocal tics of Tourette's syndrome is slowly being abandoned as well. While the impact of dopamine-reducing halo-peridol played an essential role in making the case for a neurological rather than psychoanalytic basis for the condition, the side effects were soon apparent too. By the late 1970s, many parents were complaining it wasn't a miracle drug after all, but had a deadening effect on their children. One mother described the dilemma: 'squelching a child's tics completely, to make it possible for us to deny he had a problem, or for him to stay in public school or be on a team or in a club, is as ghastly as child-beating – it's only more subtle and more sophisticated.'[10] While drugs continue to be prescribed, the emphasis (as with stuttering) is increasingly on behavioural therapies. Habit reversal therapy, much like stuttering modification therapy, teaches an individual to identify their tics in detail, increase their awareness of when one is about to happen, and then to find a competing response that sidesteps the tic. Exposure and response prevention exacerbates the force of a premonitory urge so that an individual learns to ride it without actually ticcing.

In the case of aphasia and dysarthria, the presence of significant brain damage or a progressive neurological disorder means treatment is almost never about cure. Although these conditions have been less vulnerable than stuttering or ticcing to misdiagnosis, this greater certainty brings pitfalls of its own. In the late nineteenth century, the conviction that damage to the Broca's area was both the cause of aphasia and also irreversible led to what one psychiatrist called 'nihilism': a sense that nothing could be done or was worth doing. Another doctor wrote that 'the actual therapeutic side of the question is relatively little discussed and comparatively scant attention is paid to the interest of the patient.'[11] This hasn't entirely gone away: patients sometimes complain, when they have recovered enough speech to do so, of being infantilised or ignored by the doctors who are supposed to understand them best.

Just because full recovery may be impossible, it doesn't mean immense leaps still can't be made that immeasurably improve a patient's quality of life. As with other symptoms of stroke, therapy focuses on recovering lost capability and evidence shows that through sustained practice progress can be made.[12] The ability to do this has been enhanced through cognitive neuropsychology, a data-based approach that has enabled therapists to understand the ways in which language-production is transformed by aphasia. As well as helping an individual to recover lost language skills, therapists also focus on acceptance and adjustment. In most cases, the aim isn't to recover all of the ability one once had but to find other ways of achieving good communication. Conversation training teaches family and friends to speak slowly and clearly, avoiding abrupt changes in topic and keeping background noise to a minimum. Non-verbal techniques are often introduced: pointing at a visual analogue mood scale (not dissimilar to emojis), drawing, or enhancing the use of facial expression.

As with aphasia, treatment of dysarthria is generally about mitigation rather than cure. Mostly this involves behavioural techniques, like the Lee Silverman Voice Treatment, which focus on increasing the strength and clarity of a person's voice. In more extreme cases, instrumental aids, medication (like L-DOPA) and surgical procedures may be employed. For those who have spoken fluently for most of their life, these treatments may be precious: a way of achieving continuity with their past selves. But for those born with cerebral palsy it can be a bemusing process too, with an excessive emphasis on rendering them socially acceptable rather than helping them to communicate. For Jamie Beddard, years of therapy had little impact and are lost among the other memories of childhood. 'I can't remember, to be honest,' he tells me. 'There was a lot of focus on pronunciation. I used to dribble a lot more than I do now.'

Sometimes dysarthria is linked to a neurodegenerative disorder like Parkinson's or motor neurone disease. In such cases, practising articulation techniques can feel like a losing battle. When I ask my cousin Gilly what techniques she practices in her weekly speech therapy sessions to help alleviate the dysarthria that stems from MND, she is under no illusions about the future. 'Nothing,' she says. 'It will only get worse. Energy conservation is key.' Although she can still make herself understood through speech, she is transitioning into a life that will be less dependent upon it. Gilly's speech therapy isn't really about speech at all. She works with a therapist to develop and master alternative forms of communication. Some of these are incredibly rudimentary. For instance, she shows me a set of communication cards in her bag that are coloured to signify different scenarios: yellow for travel, green for social, red for managing daily life around the town. One of the red ones reads, 'Hi, can you put the groceries in my backpack in my wheelchair after I've paid.'

Gilly's main hope lies in her AAC (augmentative and alternative communication) device. These come in many shapes and sizes and, because of advancements in digital technology, the last twenty years have been something of a golden age. Gilly uses EyeGaze, which by tracking the movement of her eyes across a screen, speaks the words that she spells. Historically, those using speech synthesisers have to choose from a limited cast of voices, like the distinctive, American-accented 'Perfect Paul' we associate with British scientist Stephen Hawking. But a new wave of voice banking technology has appeared in the last couple of years enabling those with progressive conditions to record thousands of words and phrases before their voice deteriorates, which personalises the sound of the AAC devices they increasingly depend upon. Gilly banked her voice a few years ago, but I am struck by the fact that she doesn't use it, preferring a generic Australian female voice that comes with the technology. When I ask about this, she says that her old voice isn't who she is any more. She doesn't want to hide from the present or her future. EyeGaze is part of her voice now, factory settings and all.

Although AAC devices tend to be used by those who have lost the capacity for easily comprehensible speech, it is not inconceivable – as they become more personalised and easier to use – that those with borderline disorders of speech may use them too.

In the last chapter, I showed how throughout history theories of what causes and might cure speech disorders have been ill-conceived, often damaging. Didactic treatment, the belief that there is a single and universal cure like surgery or psychoanalysis, tends to end badly. What worked best then is still what works best today: pragmatic approaches that are flexible and based on trial-and-error with an individual. Most of all, the principle of partnership between therapist and patient, listening and guiding rather than declaiming and prescribing, has proved enduringly effective, building both self-awareness and reliance in tackling the continuing difficulties a disorder may present.

While treatments that are both pragmatic and productive have taken many different shapes over the centuries, they share one ingredient: a desire, and a record of success, in building self-confidence in an individual. This isn't the general or generic confidence of the Dale Carnegie salesman, but specific: the confidence to communicate, express oneself and follow the paths one wants, even if verbal flow continues to be significantly disturbed or entirely eliminated. The primary aim of treatment shouldn't be to cure or render a person more palatable to an intolerant society, but to build or rebuild an individual's self-confidence that has often been left in tatters by years of stigmatisation and humiliation. 'You want someone you can cry with,' broadcaster Nick Robinson has said about his experience with dysphonia, 'you want someone who can listen to your fears, who believes in you, who convinces you if you work hard enough you'll get better. A huge part of a therapist's role is not just the technical and the mechanical and the medical, it's the emotional support.'[13]

In her memoir, band manager Grace Maxwell describes the extraordinary dedication of her husband's two speech therapists following a near fatal stroke:

> They learn from Edwyn and their other patients, and are continuously adding to the sum of their professional knowledge. But there is more to this dynamic

duo than any of these words can convey. They are creative geniuses, fascinated and humbled by the unknowable nature of the brain and the spirit of its possessors. This really is the key thing. It's impossible to predict the path recovery will take. There are as many routes as there are human faces.[14]

The personality of the therapist and their relationship with an individual is therefore of equal, if not greater, importance than any of the techniques they wield, which are prone to faddism. As with any form of therapy, the right practitioner can transform a person's life as much as a bad one can hamper it, by coaxing an air of optimism or despair in their client. Many 'cures' or treatments of the past, which have been revealed as bogus or empty, were immensely fruitful at the time simply because the practice of them created a sense of optimism.

Earlier, we saw how Lionel Logue helped King George VI to greater fluency, but his actual techniques, like 'three word breaks', are an ungainly form of speech modification therapy and pale in effectiveness compared to his willingness to stand beside King George VI and smile encouragingly during every broadcast. Yet this is an unsatisfactory kind of therapy, creating a sense of confidence not in the self, but in a guru figure who can guide one through difficult situations. Nicholas Mosley was also treated by Logue. 'He taught me to speak in cadences so that I could declaim like a politician in front of an audience,' Mosley recalled. 'I could do this quite well: then, when I was not with him, I would stammer as before.'[15] It seems the further one was from Logue himself the more one's speech diminished. A good therapist is somebody who guides an individual to both confidence and self-reliance and, in the process, renders themselves unnecessary: the opposite, in other words, of those celebrity therapists whose fame depends upon their enduring indispensability to their clients.

In many cases, greater fluency is a by-product of increased confidence, but not always. Either way, fluency comes to matter less to an individual. 'I had a lot of speech therapy,' says Patrick Campbell, co-author of *Stammering Pride and Prejudice*, 'but what I've found most empowering lately is just seeing stuttering as my voice and part of who I am rather than something to be fixed.' Jess Thom describes a similar change in her attitude to Tourettes. 'It was developing the language and confidence to start explaining my experiences to other people,' she says. 'That's what's been transformative – more so than any other intervention.' Comedian Lee Ridley uses similar language: 'I'm a lot more at ease with using my talker. In the past, it used to stress me out when I had to speak to new people because I knew they wouldn't know what to expect ... But now I feel differently. I don't care if I take a bit longer to reply ... I just feel more confident in myself as a whole.'[16] For me, the surge in confidence that I acquired from the Michael Palin Centre for Stammering Children in 1987 and the City Lit course for Interiorised Stammering in 2009 contributed to ensuing periods of greater fluency. But it's a particular sort of confidence: that which comes from self-knowledge, understanding what makes me tick, rather than just feeling good about myself.

The cultivation of acceptance and self-confidence is now recognised as more important than the imperative to fluency, but knowing this doesn't make it easy and the number of people who genuinely reach a place of acceptance about a speech disorder are still few. The problem is that the odds are stacked against those with speech disorders by the sheer scale of social prejudice. Even if they have the potential to overcome their own psychological barriers, they still face the daily discrimination – sometimes overt and bullying; sometimes polite and unspoken but equally debilitating – of peers, employers and families. And then there is the cultural ideal of fluency permeating all our values about language and speech which such individuals have to (imperfectly) navigate their way through each day. 'People are very judgemental,' my cousin Gilly says when I ask what she would most like to change about her experience of dysarthria. 'So the minute you are in a wheelchair or you open your mouth and you don't sound right, you're done. And your intellect is judged. And that will only get worse as my voice gets worse. I want society to be more accepting, more open and not so ready to label everyone.'

It follows that if we can prevent society from stigmatising speech disorders, then half the struggle is solved. But how does one go about changing the world? It may feel like an impossible task, but of all the options it may be the best we have. After all, over two thousand years of medical enquiry have created as many dead ends and even dangerous miscalculations (like the surgery of Dieffenbach) as they have effective mitigating tactics. While such enquiry needs to continue, and deserves support, it is not enough simply to wait for a magic cure bearing in mind how little we still know about the workings of these disorders and how resilient they have all proved against the best efforts to eliminate them. The priority is to empower individuals to experience them more positively and develop alternative forms of communication. Nobody can do this on their own: they need a receptive environment to achieve it.

It may not be easy, but we need to cultivate a new way of thinking about speech disorders within our society and culture. Not as something whose origin must be uncovered; a cure found. But something to be understood, accepted – even celebrated. The greatest hope for this lies not so much in speech therapy, which can only tackle the inner life and behaviour of an individual, but in neurodiversity: a conceptual and social movement which has rapidly gathered momentum over the last two decades.

7

Extraordinary Minds

In the late 1960s, a young British neurologist called Oliver Sacks began a long residency at Beth Abraham Hospital's chronic-care facility in the Bronx. New York was the perfect place for a doctor with an interest in the human brain: a vast metropolis with a seemingly endless supply of strange and fascinating characters. At Beth Abraham, he encountered patients with extreme symptoms of 'sleeping sickness', aphasia, amnesia and Parkinson's disease, as well as rare conditions with names like agnosia, hemispatial neglect and somatoparaphrenia.[1] Many of these individuals experienced great suffering, others were barely aware of their difference from others, all of them were unique.

Sacks quickly realised how inadequate the prognoses and treatments of the past had proven by the simple fact that such an extraordinary range of human beings were lumped together, as they were in many countries, in hospital wards cut off from the outside world. Often they were considered lost causes and received only the most perfunctory attention from the staff. Because of the overpowering influence of psychoanalysis, understanding of many neurological conditions had scarcely progressed since the 1890s. But seventy years after Charcot and de la Tourette, Sacks and others of his generation were starting to look afresh at conditions which psychoanalysis had ultimately failed to make sense of.

More often than not, these conditions affected speech. For instance, Sacks observed how patients with extreme symptoms of autism were deemed incapable of communication. Passing doctors would try and get through using the same verbal language they used with children and then give up. The more Sacks studied these patients, the more he realised they were communicating the whole time: not with words, but gestures and nonverbal expressions.[2] Crossing New York to and from work each day, he began to notice people with Tourette's syndrome not just in the hospital wards, but on the streets of the city. He realised not only how prevalent it is but the extent to which many of them managed to maintain normal lives. Following the lead of Arthur and Elaine Shapiro, Sacks tried one of his patients on haloperidol. Both doctor and patient were delighted at how effective it seemed. But, to Sacks's surprise, 'Witty Ticcy Ray', as he called him, felt increasingly conflicted about this 'cure': he complained his speech was less quick-witted, less funny, even his dreams were 'straight wish-fulfilment with none of the elaborations, the extravaganzas, of Tourette's'.[3] In the end, he settled on a compromise, spending the working week on haloperidol, but letting his Tourette's fly at weekends.

On another occasion, Sacks found a group of patients with aphasia in the ward roaring with laughter while watching the slick, hyper-fluent patter of a politician on television. At first he struggled to understand what they found so amusing: after all, this was a level of fluency they could only dream about. Then he realised it was the emptiness of the politician's rhetoric that was making them laugh. 'In this, then, lies their power of understanding,' Sacks wrote. 'Understanding, without words, what is authentic or inauthentic. Thus it was the grimaces, the histrionisms, the false gestures and, above all, the false tones and cadences of the voice, which rang false for these wordless but immensely sensitive patients. It was to these (for them) most glaring, even grotesque, incongruities and improprieties that my aphasiac patients responded, undeceived and undeceivable by words.'[4]

For the young Sacks this was a surprising revelation, but I believe people with speech disorders have always been well placed to expose the tendency to glibness, occasional deceitfulness and intolerance of other modes of communication that is inherent to hyper-fluency. They do so because they have a fundamentally different relationship to language than those who consider themselves fluent. It is still a tool for communication, but one that is inherently unreliable and requires constant vigilance to avoid tripping them up. By looking at their own speech in this way, they become aware of the way others use and misuse language.

The more time Sacks spent with his patients, the more he became convinced they were misunderstood. Historically, their conditions were always described in negative terms, as displaying a 'deficit' of ability. But what, Sacks wondered, if this was only part of the story? What if their conditions were also 'ebullient' or 'productive' in character? This question instigated a fundamental change in how we understand neurological disorders. He asked it again and again in his books, for instance in *The Man Who Mistook His Wife For a Hat* (1985) in which the life stories of his patients are offered up as demonstrations of the richness of human experience. 'Defects, disorders, diseases can play a paradoxical role,' he wrote, 'by bringing out latent powers, developments, evolutions, forms of life, that might never be seen, or even be imaginable, in their absence.'[5]

Over the ensuing decades, an increasing number of doctors and patients joined Sacks in a new celebration of neurological difference. Such conditions could be limiting, frustrating or painful, but these deficits were often partially compensated by insights and capabilities denied to others. Gradually, but with increasing momentum, the narrative around certain conditions began to shift, and it is here that I find the best hope of alleviating the suffering that accompanies speech disorders in our society. These changes occurred quickest and most demonstrably around autism and dyslexia. While not speech disorders, they impact language and provide an example and precedent by which we might think about speech in new ways.

Autism, a condition in which people experience the world and communicate differently to others, is said to affect around one in a hundred people. It was first 'discovered' (that is: identified and named) as recently as the 1940s.[6] This was at the height of psychoanalysis and for a long time it was widely believed to be caused by 'refrigerator mothers' who neglected their babies' emotional needs, turning a child in on itself.[7] As a result, the parents of autistic children experienced appalling guilt, while those with autism were seen as stunted and half-formed individuals. But in the 1970s, evidence increasingly compelled a neurological rather than psychoanalytic interpretation of autism. Neuroimaging seems to confirm this, suggesting something fundamentally different in the make-up of the brain.[8] This, coupled with an awareness of just how many people have some form of autism, has resulted in a shift in treatment. Rather than trying to force individuals into a 'normal' life, we increasingly try to understand the way they engage with the world and provide them with the tools to navigate society more easily.

'I think in pictures,' writes Temple Grandin, a world famous professor of animal science who has also provided invaluable insights into her experience with autism. 'Words are like a second language to me ... When somebody speaks to me, his words are instantly translated into pictures. Language-based thinkers often find this phenomenon difficult to understand.'[9] In 2009, American blogger Amanda Baggs posted 'In My Language', a short film using images, text and voice over, which attempts to show how she experiences the world. Rather than being trapped within herself, as people assume, she simply interacts with her surroundings more through her senses than speech. 'The thinking of people like me is only taken seriously if we learn your language,' she says. 'It is only when I type something in your language that you refer to me as having communication.' Parents of autistic children often describe breakthrough moments where they suddenly discover a new way of communicating. A famous example is that of Owen Suskind who as a child seemed unable to speak but had an obsessive interest in Disney movies.

He and his family began to communicate using dialogue from those films in the roles of animated characters.[10]

Over the decades, ever greater recognition has been given to the unique talents that people with autism display. Occasionally, we hear stories of those who perform seemingly super-human feats of memory and cognitive processing: playing an entire Tchaikovsky concerto after one hearing or learning to speak Icelandic fluently in a week. These are inspiring if extreme examples. More prosaically, evidence shows that everyday jobs built around systems rather than human interaction lend themselves well to autistic thinking. The popular stereotype of 'on the spectrum' computer programmers is not as lazy as might seem: there is a greater percentage of people with autism spectrum disorders living around Silicon Valley than anywhere else in the United States.[11] Their contribution to emerging technologies has benefited us all and is testament to a growing conviction that we need to enable those with autism to find their niche. We hear a lot about the importance of presentation and communication skills in the twenty-first century, but the world turns just as much, if not more, on the strength of those who understand systems, algorithms and patterns better than the erratic behaviour of human beings.

Autism primarily impacts social interaction and communication. Dyslexia, on the other hand, impacts reading and writing ability. It affects more people than almost any other neurological condition (according to the British Dyslexia Association around 10 per cent of the population is dyslexic; 4 per cent severely so). It was once, and still is, confused for stupidity, but it is increasingly recognised as a difference in information processing rather than intelligence. Since the 1970s, it has been carefully repositioned in public discourse.

'Children with any form of dyslexia are not "dumb" or "stubborn,"' writes Dr Maryanne Wolf, Director of the Centre of Reading and Language Research at Tufts University, Massachusetts.[12] 'Their brain is differently organised for written language. Brain imaging research suggests that some people with dyslexia appear to have a very strong right hemisphere that appears atypically activated when they read.' Since reading depends on left hemisphere activity, this explains why it takes them longer to read. However, it also explains the tendency for creative and original thinking, which is associated with right hemisphere activity. Wolf describes how, 'Inventors and artists like Thomas Edison, Leonardo da Vinci, and Pablo Picasso; modern day entrepreneurs like Charles Schwab, Rt. Hon. Michael Heseltine and Richard Branson; actors and writers like Johnny Depp, Keira Knightley, and the late Agatha Christie; all had histories of dyslexia.'

Debris Stevenson is a contemporary artist from London who works across different art forms. She is a poet, playwright, actor and dancer: skills she occasionally brings together as in her play, *Poet in da Corner,* which she both wrote and starred in. She is also profoundly dyslexic. I'm interested in how dyslexia has shaped her use of language and whether there are any lessons to be drawn for how we rethink speech disorders. When we meet, she describes the difficulties it has given her: not just around reading and writing, but also how she speaks and listens. 'If things don't have really strong context I don't understand them,' she says, speaking with rapid-fire speed and pronunciation. 'So often I'll be in a position where I have no idea what someone is talking about, but if they go back to basics and really build the foundation of what they're talking about then I'm fine.'

At the same time, Debris believes there is something intrinsically creative about her condition, beginning with a unique mental spatiality: 'Thinking three-dimensionally is a big part of it and that can be really overwhelming. I keep A3 notebooks and write in eight different colours in eight different directions and that's quite a good visual idea of what my brain is like. Something that helped me is touch-typing. It's enabled me to get as much out as possible. I can touch-type while having a conversation with you about something else.' Debris thinks the multi-dimensionality of her dyslexia is why she works across so many art-forms, but also why she works the way she does within each art form. 'In a good poem, every word has the utmost meaning,' she says. 'The rhythm of the sentence is saying something. The verbs I've chosen in context say something. The way that I say it out loud says something. Me moving says something. If you've seen me doing a poem that's the easiest way to see how I think because a poem is 3D, it's not a sentence.'

While a positive shift in perception of dyslexia is far from universal, there is far greater awareness in the public mind of both its unique qualities and the fact that it does not signify a lesser intelligence. In turn, there is less stigma attached to it than there once was. Schools are more effective at spotting it in children and supporting their needs, although there is still a long way to go. There is even a certain glamour attached to the notion of being dyslexic, fuelled by the growing list of extraordinarily creative and entrepreneurial people associated with it. When Apple ran its famous 'think different' campaign at the turn of the century, a disproportionate number of the exceptional individuals it celebrated (and implicitly linked to its own brand) are or were believed to be dyslexic: Pablo Picasso, Albert Einstein, Richard Branson and John Lennon. And it has been repeatedly claimed that Steve Jobs, the man behind the campaign, was dyslexic himself.

It may be early days still, but those changes to the way we think about autism and dyslexia since the 1970s have been seismic and continue to gather momentum, impacting how other neurological conditions are perceived. In 1998, this paradigm shift was given a name: neurodiversity. The term was coined by an Australian sociologist called Judy Singer,[13] although Oliver Sacks is rightly seen as 'the godfather' of the movement.[14] Neurodiversity describes a way of looking at neurological conditions like any other form of human difference, acknowledging the difficulties it may present to an individual but also the unique insights and abilities it gives them.

According to theorists of neurodiversity, this way of seeing is neither an abstract, academic concern nor woolly political correctness. It is urgent. We live in a diagnosis culture: the number of listed psychiatric conditions trebled in the second half of the twentieth century. Today, the NHS claims that one in four adults and one in ten children experience mental illness.[15] These statistics throw down a gauntlet: either we consider a vast and ever-growing percentage of the population as ill, or we accept that many neurological conditions may not be diseases or disorders at all. This means trying to understand as much as medicating them, celebrating their productive qualities as well as their negative ones. Most importantly, it is a 'neurotypical' majority that needs to change to accommodate a neurodiverse population rather than the other way round.

The beauty of neurodiversity is that simply subscribing to the concept is the trigger for change. An individual who begins to view their condition or disability as a unique and essential variation on societal norms becomes empowered to demand the opportunities previously denied them. Often this rejection of discrimination is enough to convince neurotypical individuals and organisations, whose prejudice may have been rooted in ignorance rather than malice, that change is necessary. Like a virtuous circle, this in turn further empowers neurodiverse communities, which leads to greater social awareness.

Although over twenty years old, neurodiversity is still a new way of thinking, and the range of conditions it encompasses has varied and expanded. Judy Singer introduced the term specifically around autism, based on her experience as the mother of an autistic child. But over the ensuing years, neurodiverse accounts of dyslexia, ADHD (Attention deficit hyperactivity disorder), mood and anxiety disorders and schizophrenia have followed in quick succession. A core argument is that such terms are all culturally determined, much as we have seen speech disorders to be. 'Whether you are regarded as disabled or gifted depends largely on when and where you were born,' writes Thomas Armstrong, author of *The Power of Neurodiversity*. 'Instead of regarding traditionally pathologised populations as disabled or disordered, the emphasis in neurodiversity is placed on *differences*.'[16]

Neurodiversity has much in common with another challenge to mainstream perception, which emerged at the same time but has focused on physical disability. In 1975, a small activist group in the UK called the Union of the Physically Impaired Against Segregation released a statement that tried to upturn several thousand years of discrimination. 'In our view it is society which disables physically impaired people,' it announced. 'Disability is something imposed on top of our impairments by the way we are unnecessarily isolated and excluded from full participation in society.'[17] The disabled academic Michael Oliver coined the phrase 'the social model of disability' to describe this conceptual shift.[18] It is a fundamental inversion of the way many people see disability and, as a result, those encountering it for the first time can struggle to get their heads around it.

One example often used to explain the social model of disability is that of a wheelchair user. Despite popular perception, a wheelchair user may be incredibly mobile, able to turn and move faster than many 'able-bodied' people. The problem arises when you put an obstacle in their way, like a flight of stairs or a bus without a ramp or a curb that is too high. All these are things produced by a society that is prejudiced towards a single, dominant type of mobility. They are creations which, in enabling an able-bodied person to do something, exclude a wheelchair user. The social model of disability argues that it should not be beholden to the disabled individual to adapt themselves to an intolerant society, but for society to adapt itself to the millions of people within its entirety who it renders disabled. It was, and is, an important stepping stone in the development of 'neurodiversity', which proposes an equally radical reframing of neurological difference.

The internet, particularly social media, has both fuelled and created a tipping point for neurodiverse and social model activism. This is partly because it allows individuals who are often isolated in their immediate environment to form micro-communities with other individuals across the world. But it is also because the form itself has opened up new opportunities for people who struggle, as so many neurodiverse and people with disabilities do, with neurotypical forms of communication. For many autistic people, language itself is not a problem, but the way human beings use it: they understand what words mean perfectly, but struggle to 'read' the facial expressions, body language and emotional subtext that provide a counterpoint in their usage. Sarcasm, for instance, is a technique of tone and manner that undermines, even contradicts, the words it accompanies. Not only does the internet put a greater focus on written language, it also empowers other forms of communication like memes, emojis, Instagram photos and short-form video that neurodiverse or disabled people may be more comfortable using.

In fact, the internet can be rightly called the first truly neurodiverse form of communication in human history, which is hardly surprising considering the role that neurodiverse people have played in designing and expanding digital technology. In a world where everybody is communicating through their computers and phones, the use of AAC devices by those who have disordered speech or are non-verbal no longer stands out as it once did. In my childhood, Stephen Hawking and his synthesised voice felt like a figure from the science-fiction programmes that I watched on television. Today, a comic act like Lost Voice Guy, who delivers his routine through his AAC, is just somebody doing what we all do: communicating first-and-foremost through the aid of a device. He just happens not to use it in tandem with vocal speech.

While far from complete, the success of these social movements is evident in the culture around us. The size of the audience watching the last Paralympics on television would have been unthinkable fifty years ago, while popular series like *Homeland* or *Grey's Anatomy* have central characters who are bipolar and dyslexic respectively. Hollywood has played an under-appreciated role too. Whatever the accuracy of iconic films like *Rain Man* (1988), *My Left Foot* (1989) or *A Beautiful Mind* (2001) in representing the conditions or characters they claim to display, their success in the box office has softened attitudes to autism, cerebral palsy and schizophrenia. This increased awareness has impacted our legal system and company policies – in particular, the Disability Discrimination Act of 1995 – making it far easier to call out and rectify discrimination when it occurs.

As theories, both neurodiversity and the social model of disability have their critics. One common complaint levelled at both is that they ignore the genuine pain many people with disabilities and neurological conditions experience. Instead, they are looked at through the brightly-lit, warm-coloured filters of the films they inspired. As a result, the medical model of disability, which sees them as problems to be treated, remains dominant. They are not, however, mutually exclusive. Both medicine and society have a role to play in alleviating the suffering of the disabled individual.

There are also huge gaps still in the achievements of disability activism. One of them is speech disorders. Discourse and debate around neurodiversity rarely acknowledges speech disorders as a category unless they are symptoms of a broader condition like cerebral palsy. Popular understanding and prejudice against stuttering and aphasia has scarcely changed since Oliver Sacks's day. This is surprising considering the central role that language and its disorders played in his writing. It may be because the movement is still developing. At the end of *The Power of Neurodiversity* (2011), author Thomas Armstrong speculates on the future of the term. 'We should probably make our definition of it as inclusive as possible,' he writes. The next wave, he predicts, will be around 'dyspraxia, Tourette's syndrome, nonverbal learning disabilities, and speech and language disorders.' Clearly, rehabilitating these conditions within society has felt less urgent and they are picked out almost as an afterthought. While this may simply reflect the vast number of people who have conditions like autism and dyslexia and the scale of discrimination against them, there are other reasons to consider too.

Earlier, I described a 'peculiarity' (to use Erving Goffman's description) about speech disorders that makes those who have them less likely to organise themselves compared to people with other forms of disability. It may have something to do with the way many speech disorders, unlike other disabilities, can be partially or even completely concealed – even if it involves reshaping every aspect of a life (career, relationships and hobbies). This ability to pass as neurotypical means that most with speech disorders don't even think of themselves as being disabled and it rarely occurs to them that the curtailing of experience is in itself a form of disability. And it also results in the 'habit of secrecy, of feeling himself cut off from his kindred; of brooding over his thoughts, of fancying himself under a mysterious curse' that James Hunt described one hundred and fifty years ago.

Such secretive, loner behaviour is hardly compatible with activism, which depends upon collaboration and outspokenness: an outspokenness that many with speech disorders feel physically incapable of performing or have learned to avoid. Finally, there is the modern tendency to compartmentalise speech disorders into distinct and self-contained silos, a differentiation that is not made in other societies at other times and discourages collaboration. By even talking about people who pathologically stutter, tic, struggle to articulate or translate thoughts into speech as sharing a problematic relationship with language, we run against contemporary practice that resists drawing comparisons across these speech disorders.

Yet this 'peculiarity' must be overcome. Since the causes of many speech disorders remain murky or poorly understood, and the long-held hopes for a 'cure' as elusive as ever, the achievements of neurodiversity and social model activism present the most promising, if not the only, opportunity for achieving lasting improvement in the well-being of those with such conditions. This is not a modest or pessimistic pursuit: remove the stigma against speech disorders and half the trouble goes. Those with borderline conditions will find their speech scarcely bothers them at all, while those with more problematic ones, like extreme aphasia, will achieve greater communication with a society more willing to engage patiently with them on their own terms.

One speech disorder already benefitting from neurodiverse thinking is the vocal ticcing of echolalia and coprolalia that we associate with Tourette's syndrome. This may be because Tourette's is not a speech disorder in itself, but a broader condition that can affect the entire body; in some cases, bringing mobility aids like wheelchairs into play. Early attempts at raising awareness of the condition tended to focus on the more spectacular and disruptive symptoms of coprolalia. Documentaries like the BBC's *John's Not Mad* (1989) had the double-edged outcome of convincing a mass audience that people with TS weren't insane, but that it was also predominantly a 'cursing' disease. The writings of Oliver Sacks were (again) instrumental in shifting attention away from the cursing of Tourette's to its creativity: not just his iconic 1981 essay 'Witty Ticcy Ray', but in ensuing books like *An Anthropologist on Mars* and *Musicophilia*. It is because Tourette's has an involuntary but nevertheless productive quality that it is more easily rehabilitated within a neurodiverse perspective, in contrast to other conditions affecting speech which seem defined, like stuttering or aphasia, simply by an inability to get words out.

Jess Thom readily attributes her sense of personal liberation to neurodiversity and the social model of disability. 'I feel very clear that the moment where my life really changed was the moment that I started talking about Tourette's and giving myself space to think about it rather than push it away and hide it,' she tells me. Soon she realised the extent to which her experience of Tourette's had been determined by the perception of others.

> The Social Model of Disability has been really important to me. We live in a society where the messages we get from a very early age are that there's one way of doing things; there's one body, there's one mind. We can't really cope with fluctuations and don't even want to describe them as grey areas. If you deviate from the norm then you are wrong or broken. But that is such a lie. And it's a lie about our bodies, it's a lie about Disability.

Gradually, Thom came to realise that these social and cultural conventions, more than her own body, are what need to change. This isn't just ideology, but pragmatism too. 'At the moment there is not loads that can be done to change the experience of Tourette's from a physical point of view,' she says, 'but there is something that we can all do about the social impact immediately which is about increased understanding.'

While there have long been organisations that raise awareness and help people who stutter, the focus has traditionally been on support for the individual rather than a radical repositioning of how stuttering is perceived. But in the last couple of years a more defiant and radical spirit has emerged: a movement which calls itself, depending on which side of the Atlantic you are on, Stuttering Pride or Stammering Pride. Central to this new spirit was a blog called *Did I Stutter?*, started by a couple of academics in North America, which became a virtual gathering space for stuttering activists across the world. One of the founders is Joshua St Pierre, a lecturer in philosophy at the University of Alberta, Canada.

'I've stuttered my whole life,' St Pierre tells me, when we talk on the phone. 'For the majority of that time I had incredible shame about my speech. And I would always assume that if communication would "break down" it was my fault because I was the disabled, negative subject.' The turning point came in the summer of 2012.

> I was working a summer job, and I'd been struggling hard to get out a sentence with this guy I was working with. I was stuttering and it came out really slowly but still clearly. In response he said, 'huh?'. So I said it again. And he said 'huh?'. Suddenly I realised that it wasn't my fault in this case. He was the one being the poor interlocutor. And I felt anger in a way I'd never felt before.

St Pierre became convinced that although he stuttered, the breakdowns in communication he experienced were more a consequence of people unwilling to listen or take the extra time to pay attention to stuttered speech. This is when he became interested in the social model of disability. 'I realised the first wave of disability activism was based on physical disabilities and those who could claim cognitive parity,' he says. 'People who could say we're just like you and therefore we deserve the same rights you do. But there wasn't any activism yet for speech disorders.' St Pierre teamed up with a colleague called Zahari Richter, who also stuttered. 'The idea of having a blank slate enabled us in some ways to decide what we wanted this to be. We took a fairly radical stance as far as disability studies go in that we didn't just want inclusion, but we wanted to call out the imperative to be fluent across society and how that affects a wide range of speakers.' They were soon joined by an American poet called Erin Schick, whose performance of an activist piece called *Honest Speech* had recently gone viral on YouTube.

In 2014, the three launched *Did I Stutter?* which became a spearhead for the newly emerging Stuttering Pride movement. From the start, the traditional aims of therapy and treatment were deemed inadequate. 'Inclusion or acceptance is a big buzzword in speech-language pathology,' St Pierre says. 'But I think that's just a watery, weak goal, because we're being asked to accept ourselves on the terms of the medical, ableist world. I think we can do better than acceptance. I think there's room for a disfluent pride. There are huge variations in how people communicate and speak and this is just one of those variations.' Through the blog and social media, St Pierre, Richter and Schick came into contact with thousands who felt the same way: people who didn't want to be accepted by society, but to change it. When I first encountered their work, it was intensely exciting to hear convictions I had felt but never shared with anyone expressed in the public realm.

Bizarrely, the first line of resistance came not from the prejudiced and discriminating masses, but from speech therapists who worried their views were too extreme and uncompromising. 'It became an issue of challenging the authority of who gets to speak the truth about the stuttering body,' St Pierre says. 'Speech-language pathologists and geneticists are the ones who have the authority to speak the truth about what stuttering actually is and therefore how we should go about dealing with it.' You can hear this in action on two editions of *Stuttertalk*, a podcast presented by speech-language pathologist Peter Reitzes, from 2014. Over the last ten years, *Stuttertalk* has been hugely important in raising awareness about the nuances and issues around stuttering and building an online community. But confronted by *Did I Stutter?*, Reitzes gets audibly queasy. The whole social model of disability is one he admits he's never considered before, but he is open-minded. Then the controversial issue of 'informed consent' comes up.

According to St Pierre and Richter, children should not be submitted to speech therapy because they are too young to comprehend what they are really getting into. Needless to say, this is something Reitzes, as a therapist who works with children, cannot abide. Therapy, he insists, has progressed: it no longer reinforces stigma, but gently helps children to both understand their stutter and develop some management tactics if they choose to do so. But St Pierre disputes whether this is possible: once a child is in therapy they are trapped in a perception that their speech is a problem that must be dealt with. What's more, this is happening at such an early age it is likely to determine how they view themselves for the rest of their lives, a phenomenon that Wendell Johnson had identified decades before. St Pierre argues that until an individual can make a decision to enter therapy for themselves it should not be thrust upon them. The issue of 'informed consent' becomes a line that neither side can compromise on.

'Our position has always been that it's every person's right to go to speech therapy,' St Pierre says.

> We don't have anything against speech therapy in itself because we understand that we live in a shitty, ableist world that makes it hard for people. Our problem with speech therapy is that we don't think there's actual, genuine choice. Speech therapy is seen as a necessity because fluency is seen as a necessity and we don't make space for other ways of communicating. In lots of ways it becomes compulsory even if it isn't actually ever said. It's an imperative if you want to be happy. If you want to have a good life you have to go to speech therapy and fix your speech. And I did that for years and years. I had been trapped in this world of self-hate for my whole life and then I was liberated by disability activism. It isn't just this heavy thing I do: it's changed me, I'm a different person. We want other people to have the option to experience this too.

The issue of consent is a difficult issue to resolve, and ultimately a personal one. I benefitted immensely from speech therapy as a child, because I was lucky enough to be among the first intake of what would become the revered Michael Palin Centre for Stammering Children. I can also see how the wrong therapist, or at least a bad dynamic with one, might have exacerbated my stutter. I think a lot about how I will respond if my own children begin to stutter. If they do, my approach will be to wait until I am absolutely sure it has become a problem for them personally before seeking help. Even in this outcome, the important thing is to resist unquestioning faith in the authority of a therapist but monitor closely the dynamic they create with my child; prepared, if need be, to change therapist or desist entirely.

The impact of *Did I Stutter?* has been in the discussion around stuttering, largely online. It's no coincidence that the growth of Stuttering Pride, as with so many identity movements, has developed alongside the emergence of social media. 'Before *Did I Stutter?*,' Joshua says, 'there wasn't much critical dialogue that was happening, but I've been in contact with tons of people, little activist communities are popping up everywhere.' And here's the double bind: social media, which has done so much to promote hyper-fluency, fake news and trolling, has also enabled the group formation and speech activism that Erving Goffman once considered impossible. It's a reminder that technology in itself is rarely to blame, but the uses it is put to. The imperative to organise becomes greater than ever in order to outweigh the spirit of intolerance that governs so much social media usage.

In the last couple of years, Stammering Pride has emerged in the United Kingdom. One particularly dynamic cohort includes therapists and alumni of the City Lit adult speech therapy department in London, who are applying a neurodiverse and social model of disability theories to stuttering. Patrick Campbell, co-editor of *Stammering Pride and Prejudice: Difference not Defect*,[19] is central to this movement. 'I no longer believe in all these speech techniques to try and improve fluency,' he tells me when we meet in a Manchester café.

> I think that almost hinders people who stammer because it just encourages the thought that they shouldn't be stammering. It all links back to the stigma which came to stammering over the years: the idea that people who stammer are less competent, less able than other people. And this seems ingrained in our whole culture. People who stutter on TV or film are always the bad guy or stupid. As a child who stammers in that environment you take on those views of yourself, you become self-stigmatised to think those things of yourself. You think you're less able, less competent because that's what society has been telling you throughout your life.

Changing that wider perception of stuttering is an arduous process of day-in day-out activism. Even as we talk, Campbell confronts me on the adjectives I use to describe stuttering. 'You used the term "flared up" when you said you had a bad stammering period which is like saying it's got worse,' he says. 'But why did you use that term? Why did you not say "stammer more"? "It's got worse" implies there's a subjective value judgement against a person.' Having grown up using the binary terminology of fluency and disfluency, of normal and disordered speech, I struggle to think of my speech outside of a positive to negative spectrum. Following this encounter, I resolved never to use negative adjectives to describe anybody's speech disorder unless used by the individual in question.

The emergence of an activist movement around stuttering is immensely significant, a rebuttal of the peculiar inability to organise that Erving Goffman identified, but it is early days still and mainstream attitudes to stuttering remain fundamentally unchanged. And then there are those disorders of speech which threaten to be left behind entirely. Perception of dysarthria is tied to that of those conditions it accompanies, so while experiences of cerebral palsy have been central to the evolution of the social model of disability, one can hardly say the same about Parkinson's disease or brain damage. In each case, irrespective of how often people with such conditions say it is their speech which troubles them most, dysarthria is always seen as a by-product or symptom of a broader condition rather than one which warrants a neurodiverse analysis in its own right. Many people who have dysarthria don't even know the name for it, which makes it even harder to educate a fluent mainstream how best to respond and behave when they encounter somebody who struggles with articulation.

In some ways, this situation of semi-acknowledgment is preferable to the unremittingly bleak perception of aphasia. Many exceptional individuals have had aphasia, but their exceptionality is deemed to have preceded it. In fact, aphasia is often understood as the very thing that has deprived them of such exceptionality, tearing away abilities they took for granted and seeming to offer nothing as consolation. If it appears in the biography of a great person, it tends to be in the final chapter when a series of strokes carries them speechless to their grave. The testimonies of those who experience aphasia, and live a long life after, are often filled with frustration and despair. And since activism depends upon a degree of verbal eloquence and fire in the belly, the experience of aphasia, which is confusing, frustrating and diminishes self-confidence, is hardly conducive to changing mainstream perception. After all, how do you 'speak up' when you are lost for words?

The strength of neurodiversity is that it argues for a productive value within neurological conditions as a counterpoint to traditional notions of deficit. Yet this is also its weakness, for it creates an implicit hierarchy between conditions based on which are the most 'productive'. Neurodiversity has prioritised autism, dyslexia and, more recently, Tourette's syndrome, where such an argument can be fairly easily made with the aid of celebrity advocates and the retro-diagnosis of historical figures. But it has floundered with conditions where arguments for unappreciated productivity are hard to identify or when the suffering of individuals resists any positive spin. How can one make the case for the blocked, distorted and lost words of stuttering, dysarthria and aphasia when they seem to offer no productive qualities whatsoever?

This then is the immense challenge we face: the greatest hope for alleviating the suffering and stigma of speech disorders currently lies not with medical science, which acknowledges there is no 'cure' for such conditions, but with the dynamic movements of neurodiversity and the social model of disability. They, at least, attempt to diminish many of the worst aspects of such conditions by tackling the prejudices held by both society at large and even by individuals themselves. Yet speech disorders are rarely included in such discourse, partly because they are considered disorders rather than disabilities, partly because those who have them have been (mostly) reluctant to organise. But most of all because the argument for their productive or ebullient nature has, with the exception of vocal ticcing, been hard to make.

In the following chapters, I present the case for the productive qualities concealed within speech disorders. Despite the suffering they cause, they can also enrich an individual's experience of life and day-to-day communication. They inspire a creative energy that is not only productive but unique, with consistent and recurring qualities. And they play an important part in challenging some of the more dangerous and intolerant tendencies in our society.

In doing so, I aim to tie speech disorders closer to the benefits of neurodiversity, to achieve not only acceptance but appreciation too. This is not a plea for tolerance: all individuals, and all neurological disorders, deserve to be treated with dignity and respect without having to justify themselves. Yet the methods of neurodiversity are undeniably effective in speeding up the process of assimilation, alleviating stigma and, in this case, liberating us all to think about our speech in different ways. This is something everyone can benefit from. Since the rampant fluency prejudice, or hyper-fluency, in our culture has the unexpected effect of narrowing as much as expanding human experience, then learning to appreciate disfluency may enable us to speak and think in different ways. 'Diversity is always hard for us,' linguist Daniel Everett tells me. 'But new information comes from innovation or difference. If everyone talks the same, we don't think about the nature of our speech. Just a simple case of one child stuttering can cause us to learn about what speech is and what tolerance is.'

8

Virtuous Disfluency

As a child, the future George VI was ragged mercilessly by his siblings for his speech while his father, the King, watched on. Although he didn't know it, this humiliation connected him to thousands of other children growing up around the country. One was Aneurin 'Nye' Bevan: two years younger, born in the Welsh mining valleys, with a stutter to rival that of the prince. While George's was to be a life of unrivalled privilege, Bevan had little more to look forward to than one of ill health and physical suffering in the local colliery where his father worked. The idea that he would one day found the National Health Service would have seemed to him deluded fantasy. Yet when the two men became acquainted later in life, their shared experience of stuttering meant they enjoyed talking together.[1]

Like most of his classmates, Bevan left school at thirteen and went to work. The conditions were intolerable and he became increasingly aware of the injustices of society: the vast gaps between rich and poor, the strong and the weak, the healthy and the sick. 'A young miner in a South Wales colliery,' he later wrote, 'my concern was with the one practical question: Where does power lie in this particular state of Great Britain, and how can it be attained by the workers?'[2] Bevan concluded the best hope for addressing this question lay with the trade unions so he joined the local chapter of the South Wales Miners' Federation, known simply as The Fed, and became one of its most tireless and dedicated activists.

Even as a teenager, Bevan was recognised as an asset, but there was a problem: union activity turned upon speech, whether chapter meetings, conferences, rallies, or one-to-one advocacy. The ability to talk well and persuasively was the pre-eminent quality necessary for an effective trade unionist. William Abraham, one of the founders of The Fed, was an orator first and foremost, renowned for his deep booming voice and ability to switch effortlessly between English and Welsh. Bevan, on the other hand, stuttered. It broke his flow, muddled the meaning of his words, created opportunities to be interrupted or dismissed. It seemed that the fire within, the potential to change the world, would remain unharnessed.

Bevan did what few others would: he simply kept at it. He spoke up falteringly but regularly in meetings. He volunteered himself for every public speaking opportunity imaginable. He addressed groups of tired and angry miners on the importance of personal sacrifice, on the need to do more for the union, on how new rights and greater power would be claimed. In his recreational hours, he studied the dictionary and thesaurus in the library, building up a vast vocabulary so he could replace difficult words with synonyms at any given moment. He stalked the hills above the town reciting poetry against the wind to develop his voice, before returning down to address another meeting.[3] Decades later, when asked how he overcame his stutter, he replied, somewhat grimly, 'By torturing my audience.'[4]

Bevan was still a young man when he won a seat as a Labour MP in the 1929 general election. His years of activism and incendiary speaking saw him quickly emerge as one of the most formidable forces in parliament: a large, fleshy-faced man in a trademark pinstripe suit. He soon rose to the senior ranks of the Labour Party, not least because he was one of the few who could take on the great orator of the other side of the house: Winston Churchill.

Churchill, of course, had his own speech impediment. While there is continuing uncertainty about what exactly it was, experts believe it was probably a severe lisp rather than a stutter.[5] As a young man, he consulted a speech therapist and it is thought that the strange and distinctive pronunciation he became famous for ('Narzees' rather than 'Nazis') was in part a way of broaching difficult sounds. In an early essay on rhetoric, he argued that a speech impediment, rather than being a handicap, could prove 'of some assistance in securing the attention of the audience'.[6] Bevan, likewise, learned to put his stutter to good use. Once, in a particularly vicious exchange with Churchill, who tended to dismiss Bevan as 'a squalid nuisance', he leapt to his feet in the House of Commons, exclaiming, 'I welcome this opportunity of pricking this bloated bladder of lies with the ...' For a moment he stuttered on the letter 'p', while the room looked on expectantly, then released: '... poniard of truth.'[7] The MPs listened carefully to what he then had to say.

Bevan's hour came immediately after the war. When Churchill was ousted in a Labour landslide, Bevan was appointed Minister of Health in Clement Attlee's government. With a large majority behind him, this was his moment to address some of the structural inequality in society that had so hampered the mining community from which he came. Bevan pushed through the National Health Service Act in 1946, pledging free healthcare for all, and launching the NHS that is so close to British hearts today. Bevan's achievements were great and, we might conclude, happened in spite of a debilitating speech disorder. But then again, maybe it wasn't debilitating at all. Maybe he achieved what he did not in spite but because of it.

Bevan's story is one of many concerning extraordinary individuals who both have a speech disorder and also scale the heights of their chosen profession. It is inspiring because the odds seemed so against him, yet the prize – the founding of the NHS – was so great. It raises these questions: can a speech disorder be of benefit to an individual, and are there circumstances where the loss of control and struggle to speak result in better speech and more effective communication? Drawing up a list of these benefits is often used in speech therapy: rather than dwelling exclusively on the negatives, those in treatment are encouraged to identify some of the positives too. This may be difficult at first because the experience of humiliation and frustration tends to eclipse all else, but gradually some are teased out.

Such 'benefits' range from the trivial (extra time in oral examinations) to the pragmatic (a legitimate reason not to participate in public speaking exercises that fluent speakers may dread just as much) to the holistic (an enhanced sense of compassion because of one's own suffering). From the testimonies I read in books and articles, as well as the interviews I conduct, there are certain benefits that emerge repeatedly as themes. Considered together, I think they provide the basic material for a neurodiverse account of speech disorders; one which seeks to celebrate their difference rather than simply, and grudgingly, tolerating it.

One group of such benefits concerns the personality of an individual. The psychological impact of a speech disorder is vast, resulting in another set of symptoms often more debilitating than the physical ones. A personality can be shaped by a disorder, as it generally is through stuttering and tics, or it can be reshaped, as it is by aphasia and acquired dysarthria. The experience of these conditions is mostly negative: humiliation and shame lead to secretive, internalised behaviour. People with speech disorders can lack confidence and avoid careers and relationships. But these psychological traits are by no means always detrimental to an individual's well-being: there are always those who strive to compensate or to achieve despite, or rather because of, those negative feelings. As artist Brian Catling tells me, 'There's a restlessness and a sort of bloody mindedness' to his character, derived from a lifetime of struggling to get his words out.

The French revolutionary Camille Desmoulins, Aneurin Bevan and – to take a contemporary example – Joe Biden are all renowned orators who stuttered. All of them grew up believing their speech disorder would likely render them unfit for public life. No less remarkable is the way that young people who struggle with speech seem to be preternaturally drawn to acting. Rowan Atkinson, Samuel L. Jackson, Nicole Kidman, Marilyn Monroe and Bruce Willis – to take a handful from a long list – all had, or have, stutters.

'My parents took me to speech coaches and relaxation coaches,' says English actor Emily Blunt. 'It didn't work. Then one of my teachers at school had a brilliant idea and said, "Why don't you speak in an accent in our school play?" I distanced myself from me through this character, and it was so freeing that my stuttering stopped when I was onstage. It was really a miracle.'[8] Blunt is one of many who stutter in everyday speech but are inexplicably completely fluent when performing a role. The best theory I have heard for this is the one suggested to me by Dr Kate Watkins at the Oxford Centre for Human Brain Activity, which posits that recital and performance use slightly different parts of the brain, circumventing those which prompt a stutter.

The phenomenon of the actor who stutters is not a recent one but dates back as far as the 18th Century. In the 1770s, the actor, playwright and novelist Elizabeth Inchbald launched her career on the stage in spite of her stutter and the discouragement of her family. Looking back two decades later, she marvelled at her perseverance. 'It has been the destiny of the writer of this Story', she states in the Preface to her first novel, *A Simple Story* (1791), 'to be occupied throughout her life, in what has the least suited either her inclination or capacity -with an invincible impediment in her speech, it was her lot for thirteen years to gain a subsistence by public speaking.'

Despite such modesty, she enjoyed considerable success on the stage and only retired when her career as a writer became more lucrative. *A Simple Story* is one of the great novels of the late 18th Century, and she wrote twenty plays besides.[9]

The link between stuttering and acting extends to other speech disorders too. The comic actor Dan Aykroyd experienced vocal tics in his early teens. 'I had a slight touch of Tourette's,' he has said, 'which means you talk to yourself and bark and cry out at night.'[10] Dash Mihok is an American actor best known in the UK for playing Benvolio in Baz Luhrmann's *Romeo + Juliet*. Through his life, he has experienced the full range of symptoms of Tourette's including coprolalia. As a teenager, acting became a way of concealing his tics, but also what made him interested in embodying other roles. Whenever he noticed other people with tics or quirks on the subway, he would mimic them, trying to feel what tics were like for others.[11] Jamie Beddard, who has dysarthria (as a result of cerebral palsy), struggles to explain his own experiences on stage and screen. 'It was completely bizarre becoming an actor,' he says, 'because as a kid a lot of people stared at me and I tried to normalise myself, and then you suddenly become an actor where you want people to stare at you.' All these examples tell us that although most people with speech disorders avoid professions that require constant public speaking or verbal performance, there is a smaller but high-achieving set who embrace them, determined to succeed in precisely the area in which others most expect to fail.

Drive and ambition (often in precisely the one area

that is meant to be out of bounds) are frequently cited as compensatory qualities that emerge alongside fear and stigma. Another is that of empathy. The person with a speech disorder may be able to lead the life of the neuro-typical: riding the same train, going to the same office, socialising in the same places. Yet their disorder, often encountered most forcibly in their youth, gives them an experience of and insight into the perspective of other marginalised or disabled people. This ability to not just perceive but feel the vulnerability of others can seem like a special power.

'I do believe in a strange way that stammering is a gift,' writes a young woman called Felicity in *Stammering: Advice for All Ages* (2008). 'To the outside world I appear fluent, happy and confident; inside I am constantly carrying around the burden of ensuring that as few people as possible find out about the stammer I've been hiding for as long as I can remember ... I am so much more aware of other people and their feelings because of my heightened awareness of myself.'[12] According to Joe Biden, stuttering is 'the best thing that ever happened' to him. Like Bevan, he strove for greater fluency as a young man by reciting poetry (although he did so in front of his bedroom mirror with a torch before his face rather than tramping the hill-sides), becoming a better orator in the process. And it gave him the empathy that a great politician needs. 'Stuttering gave me an insight I don't think I ever would have had into other people's pain', he has said.[13] In my case, I am sure the experience of stuttering has made me a more empa-thetic person because I know how much lies beneath the surface of the words we say. I'm forever listening for the emotional subtext in even the most mundane exchanges. While this is not a quality unique to people who stutter, what capacity I have for it comes from my troubles with speech.

A wonderful quality about empathy is that it is contagious. When I ask my cousin Gilly, rather tentatively, whether there are upsides to dysarthria and motor neurone disease, she is quick to reply: 'I'm a better listener. I'm a better communicator. I don't interrupt people any more because I can't.' This has transformed her relationship with her two young children. 'I'm a better mother,' she says. 'I can't yell at the kids. I can't lose my shit. I'd love to run around like a whirlwind, but I can't. Everything is so much more planned and orchestrated. And I think they become better people for it. They have to listen to the words I can say. And then they listen at school. And they understand they can't talk over people.'

Of course, there are those speech disorders that effectively remove someone from the mainstream; a world they may have long been part of. The testimonies of people with aphasia tend to focus unremittingly, if understandably, on the frustration and moments of despair. Yet, over time, as some speech returns, certain individuals describe a particular wisdom and sense of peace caused by their experience. In *The Word Escapes Me: Voices of Aphasia,* one woman describes how her 'faith became even stronger', another describes how she 'felt inexplicable, loving empathic feelings for everyone' around her.[14] 'I came to understand that aphasia is not something the clients resent (at least not all of the time),' writes one therapist. 'On the contrary, many of them experience gratitude, not in spite of, but because of their aphasia, explaining how it has rendered them wiser, stronger, and more humane than was possible before. One client mentioned a higher level of consciousness and a sense of interconnectedness with others and with the world at large that was achieved as a result of aphasia's impediment on his ability to work.'[15]

Drive, compassion, wisdom: all these are qualities that may arise in an individual because rather than in spite of their disorder. At the same time, this does not amount to a universal rule. There are just as many individuals whose ambition is thwarted by a greater loss in confidence, who are quite understandably too consumed in their own suffering to feel much compassion for others, and who never translate the frustration of disordered speech into wisdom. They will all be different for it, though, and any workplace or social network that values diversity of experience will value the insights that a person with a speech disorder might bring to a group.

A speech disorder affects the personality of an individual in both positive and negative ways. Likewise, the impact on their use of language can be constructive as well as an impediment. People with dysarthria and aphasia may find that their speech falls short of achieving its primary goal of communication. They may increasingly depend upon augmentative and alternative communication tools like pictures, mime and gesturing as well as speech synthesisers. Because the utterance of simple words and phrases requires a great deal more energy, language becomes a precious commodity. As a result, each word has more value and potency because of the deliberation that has gone into the choosing and delivery of it: a joke can be funnier, an observation more astute. There is a precision of language not generally present in the loquacious patter we associate with hyper-fluency.

Jamie Beddard describes how cerebral palsy makes him a more efficient communicator: 'I need to be more economical in my language,' he says. 'I think about what I say more because it's more effort and I don't want to spend effort talking shit.' He continuously substitutes words that are simpler and easier to understand beneath his 'guttural' voice. 'All the time I'm looking for short cuts,' he says. 'For instance, I swear a lot more than I might do because swearing is a short cut to sentiment. Everyone understands a swear word, but I never swear when I'm writing and whenever other people swear I have a go at them.' This imperative for verbal precision makes him acutely aware of how much others might benefit from it. 'I've always wanted to test this idea that you only get a few thousand words in a day and once you've used them you have to shut up,' he says. 'I think that would make us a better society.'

My cousin Gilly agrees. 'I used to be a lot more wordy,' she says, 'but I can't be. I have no stamina so I've had to adapt. You think about what you're going to say before you say it, because you don't want to say it twice.' This change is evident in her speech patterns. She speaks in short sentences and has mostly disposed of conjunctive adverbs (like 'accordingly', 'however' and 'indeed') that connect clauses or sentences. This necessary economy of speech has given her abilities she previously didn't have. She describes her astonishment at hearing herself suddenly speak up in a book group and succinctly summarise a chapter that was causing confusion for others. 'That is not my skill,' she says. And yet, now it is.

Everyone with a speech disorder has to learn, as James Hunt claimed, 'to speak consciously as others speak unconsciously'. This should not be confused, as it so often is, with inferior linguistic competence. Rather, the reverse is true. Through struggling with speech, those with disorders become more aware of the ways in which it is used and develop compensatory tactics for managing their own. For the person who stutters, speech is far more than the utterance or performance of words you have in your head. There is a feedback loop between brain and mouth in which both send out warning signals about words or sounds which are proving or likely to prove problematic. This, in turn, allows the brain to draw upon a range of tactics and techniques to mitigate or entirely avoid the stuttering event.

Like dyslexics, many people who stutter talk about their three-dimensional or lateral way of thinking. The best metaphor I can think of is that of a slot machine: I both see and feel the sound of possible words, like the symbols on the reels, that are kept in play until the right combination locks into a place and a phrase I can utter without stuttering comes out. Everyone finds their own way of describing this. 'If I see an L or an R coming over the horizon,' says Brian Catling, 'or if I stick on one I don't even know is going to come, I will then step aside, open the door to the next room. It's a bit like a memory theatre. I find another word, an alternative word, sometimes a word I've never used before. And then I step back in the room with that word and bring it into the conversation. And when you look back you say I've never said that before, that was quite an interesting room, there's nothing in it except that word.'

Dr Clare Butler, a senior lecturer at Newcastle University, has researched stuttering in the workplace. In an interview on BBC radio, she talked about how people who stutter 'draw on a different sense of space. They use their own space in their heads to have two or three conversations on-going at the same time. They will use different words dependent on which word they can say.'[16] Dr Kate Watkins describes the difference in white matter in the brains of people who stutter, as well as the heightened activity at the moment of stuttering, suggesting a demonstrable rather than purely speculative difference in neurological activity. These differences aren't just a reflection of the compensatory energy needed to achieve what others do effortlessly, for the whole experience of language is different. To consider this simply in terms of 'impediment' or deficit is to misunderstand a relationship with language that is generally more, rather than less, productive.

'If stuttering is an impediment,' writes linguist Steven Connor, 'it is also oddly generative. Stutterers tend to become skilful synonymisers, trick-recyclists, unbelievers in the church of the mot juste.'[17] David Mitchell, the author of Cloud Atlas, has said that 'your stammer informs your relationship with language and enriches it, if only because you need more structures and vocabulary at your command'.[18] It is more than likely that the stuttering kid in the classroom who can't get their words out, who is laughed at and called an idiot, has – paradoxically

– a superior linguistic versatility than any of those laughing at them. As a kid, I fell in love with rap music even as it was still developing as an art form because of the way it plays with words. At a time when reactionary critics were dismissing the entire genre out of hand, it seemed to me endlessly inventive and the only lyrical form that came close to the way words rushed and danced through my head.

The cultural historian Marc Shell aptly uses the term 'bilingualism' to describe the way a person who stutters always translates the words they want to say into those they can say.[19] American author David Shields gives an amusing, if exaggerated example of this in his novel *Dead Languages,* when his hero wants to say in class that the American Revolution was caused by an unfair distribution of wealth. He dodges so many danger words, he ends up saying, 'The Whigs had a multiplicity of fomentations, ultimate or at least penultimate of which would have to be their prediction to be utterly discrete from colonial intervention, especially on numismatical pabulae.'[20]

Like the synonyms used by people who stutter, the vocal tics of Tourette's are an intervention in verbal flow: the difference is that they are neither chosen nor deployed but happen against an individual's will. As with stuttering, there is now evidence suggesting enhanced rather than reduced neurological activity between those who do and don't have Tourette's. 'Research examining children with disorders such as Tourette's syndrome usually explore difficulties or weaknesses,' wrote the authors of a recent study.[21] 'We wanted to examine potential areas of strength, as a way to broaden understanding of this disorder.' What they found was that children with Tourette's seem to process language faster than other children. They are quicker at assembling both the words of speech (morphology) and its sounds (phonology). Rather than representing a deficit in linguistic function, the presence of tics signals an enhancement. Jess Thom describes her own discovery of this as a Damascene revelation:

> One day my friend Matthew said to me that 'Tourette's is a crazy, language-generating machine', and told me not doing something creative with it would be wasteful. That sentence transformed how I thought about it. I have no idea why that resonated in that way. I think it was partly because I liked the idea of a machine and the idea that maybe the tics are an amazing product, they're a sort of overflow. And also because I'd been brought up believing being wasteful was very very bad.

From that moment, she saw her condition as a rare and beautiful quality as well as a disability:

> The thing I most value about Tourette's is that my tics will often draw attention to the details in the world around us that I would never otherwise recognise. For instance, I've got a really surreal and strange relationship with the lamppost that I can see from my bedroom window. Every night when I go to sleep, I brush my teeth, I put on my pyjamas as I get into bed and then I shout at the lamppost until I fall asleep. And there's no rhyme or reason for that other than

my tics draw attention to these things in the world that I wouldn't normally notice. For whatever reason or however randomly, I feel really lucky to have that relationship with the world.

While we associate vocal tics almost exclusively with Tourette's, they can develop alongside other speech disorders. In her diaries, the eighteenth-century writer Fanny Burney shows how George III used to cry 'what? what?' at the end of sentences. 'In the King, it is a mere habit,' she wrote, 'got from a disposition to stammer, which it seems something to relieve.'[22] 'If the King laughs, all laugh' – as an old saying goes – and 'what!' became a trademark exclamation of the upper classes throughout the 19th and early 20th centuries. It even became a common greeting in the term 'what-ho!'. Although they didn't know it, it is possible all who used it were unconsciously participating in a form of stuttering modification. While the sound of a stuttered word may not in itself be creative, the tricks we develop – whether consciously or unconsciously – to get around it are.

Vocal tics are also common in forms of aphasia where words are not necessarily lost but scrambled. In *Jargonaphasia*, edited by Jason W. Brown, a number of transcripts are presented, including that of a professor whose speech takes the form of an involuntary dialogue with a somewhat undermining alter ego:

My trade? Well, I have a trade that is nearly identical to that of others to that. However, he is a professor. He is a professor. Well! It is. It is hard. It is hard for

me. It is difficult because hm I am. I am in charge
of – wait! – I am a professor. I am. I am a professor.
How can I put it? I do nothing at all. I am in charge
of seeing to it that baked clay is being conditioned for
most people ...[23]

Because such interventions in linguistic flow are either
involuntary (as with tics) or compensatory (as with synonyms), we tend to ignore any productive value they might
have. But I think this is simply because they don't fit any
of our pre-existing notions of creativity. One dominant
definition, for instance, revolves around intent: deciding
and striving to turn some idea in the mind into material
form or action. It is because it is involuntary and without
an obvious organising principle that I think jargonaphasia is named so pejoratively. In direct contrast, another
popular notion of creativity is that of the genius, pluck-
ing fully formed compositions out of thin air. To some
extent, this is involuntary, although the work itself is
deemed god-given and inspired. Tics, on the other hand,
are seen to be the refuse of a disordered brain. Yet despite
not fitting into existing theories of creativity, it seems to
me that they, along with the involuntary synonyms and
jargon that also accompany speech disorders, are extraordinarily so.

The writer David Crystal defines linguistic productivity as 'the capacity to express and understand a potentially
infinite number of utterances, made by combining sentence
elements in new ways and introducing fresh combinations
of words'.[24] Children are naturally very creative in this
regard because they are experiencing language for the first

time (twenty years ago, a popular television show called *Kids Say the Funniest Things* capitalised on this for entertainment). But, as we get older, we increasingly speak in set phrases that are part of our culture. When we say 'in a manner of speaking' we do not say the word 'in', wonder what to say next, add 'a', then decide to say 'manner' and so on; we simply pluck the phrase from a mental stockpile for deployment before launching into the next one. This is why transcribed speech often seems a string of clichés, and it is why we value speakers who can produce an original turn of phrase so highly.

Arguably, the potential for linguistic invention depends not on enabling, but on resisting flow. 'If a speaker is interrupted at a random point in a sentence,' Steven Pinker writes in *The Language Instinct*, 'there are on average about ten different words that could be inserted at that point to continue the sentence in a grammatical and meaningful way.'[25] The problem is that without such interruption the opportunity to use any but the most obvious words rarely arises. While we admire the conscientious speaker who internalises such interruption, scouring ahead even as they speak for the most effective rather than predictable words to use next, the very act of interruption – whether voluntary or involuntary – has a creative potential. It opens up the possibility of using one of the (on average) nine other words that are more rarely employed, but which might make for a more striking and original sentence. Fluency, on the other hand, is not an enabler but a blocker of such creativity.

Speech disorders are ostensibly conditions that impede the smooth, clear delivery of language. The perception, reinforced by the language used to describe them, is always of deficit: words are 'blocked', 'distorted', 'lost' or 'unwanted'. Yet the anecdotal evidence, increasingly supported by neuroscience, suggests this is precisely what makes them so productive and even (to use Oliver Sacks's term) ebullient.

The language of stuttering is defined by its ornate and sophisticated synonyms as much as its blockages; that of Tourette's is as playful and witty as it is primal or obscene; that of dysarthria is as economic and precise as it is obscured and hard to understand; while that of aphasia is as precious as anything that is hard-won, through its very absence forcing greater emphasis on other forms of communication that we may have long neglected. After all, if we were as skilful at hum speech as the Pirahã people of the Amazon reportedly are, then a problem with words would be of far less import than it currently is, for our dependency on them would be smaller. All these disorders create regular interruptions in the flow of language, opening up a space of creative potential to say or do something unexpected.

It isn't only a subject's speech that is both impeded and enriched by a disorder. Conversation requires two or more people, and so it is communication itself that is impacted. Generally, this is viewed in negative terms – as disruptive, time-consuming and a source of awkwardness. The more advanced a speech disorder is, the more other interlocutors are obliged to adapt. For anyone not used to disfluent speech this can be disconcerting, even nerve-racking at first, but it is also productive. 'Stuttering breaks down normal modes of communication,' activist Patrick

Campbell tells me. 'It can make conversations which are normally superficial suddenly very deep. If you stammer when you're saying "how are you?" it shows you value who they are as you have to put more effort into what you're saying. It's not just a superficial question to you, it's something you have to push through. Stuttering adds value to almost every word you say.' The American writer Darcey Steinke describes it powerfully: 'Stuttering is a violent incantation that can break open normal conversation. What happens in that breach is up to the stutterer and her listener.'[26]

Recent history shows that the case for difference rather than deficiency tends to stem from activists within a neurodiverse community. In the case of speech disorders, fluent speakers will continue to be disconcerted and deride stuttering, tics, aphasia and dysarthria unless a counter case is presented to them. Much of this will depend upon showing, through example and argument, not only that there are productive qualities as well as deficits to such conditions, but that the prejudices of a hyper-fluent society, which recognises only one way of communicating, are flawed. This isn't just about raising awareness of disorders, but showing how much is lost when we ignore them.

Popular perception imagines a sliding scale in which the closer one is to exemplary fluency, the better and more fulfilling one's attempts at communication are likely to be. Yet sometimes the reverse is the case. Jonathan Bryan is a teenager who has cerebral palsy and can only communicate with his eyes. In his memoir, written with the aid of a spelling board, he describes the near-transcendental experiences he shared with another boy who is also 'non-verbal'. 'Friends have always been important to me, but my relationship with Will is unique… Bonded by our similar disabilities, we have never needed words; instead we look deeply into each other's eyes and together we disappear into our fantasies.'[27] On one occasion, such a moment of transport is torn back to earth by the clumsy interruption of a carer. 'Looking into his [Will's] piercing blue eyes we connected at a level beyond words,' he writes. 'Together we travelled the landscapes of our imaginations; outwardly vacant, inwardly amusing ourselves, until our journey was abruptly interrupted. "Hello, Jonathan," sang a voice dripping with enthusiasm. "How are you today? Hello, Jonathan, are you here today?"' One of the reasons Bryan wanted to master the spelling board was so he could complain about the endless disruptive and infantilising interventions from well-meaning but unconsciously prejudiced social workers.

The case against hyper-fluency (for that is what the case for speech disorders depends upon) may not be so hard to make. Increasingly, there is something in ourselves that acknowledges its limitations. I think this explains why the response to Marina Abramović's non-verbal performance art has been so ecstatic. Not only *The Artist is Present* in New York's Museum of Modern Art, but *512 Hours* at the Serpentine Gallery in London. In the latter work, Abramović was simply present in the gallery space for 512 hours, occasionally directing individuals in meditation, or leading them through the rooms, but largely in complete silence. Over 130,000 people visited the show – a large number for a small gallery.

The experiences of Jonathan Bryan and Marina Abramović are a world apart, but they reveal in different ways how little our ability to communicate is dependent upon speech, and how enriching it can be to allow space for other types of communication. In contrast, the symptoms of speech disorders may seem quite mild: a slight stutter that repeatedly trips up a speaker and forces a listener to wait or interrupt less; the vocal tics of Tourette's syndrome that require a disentangling of involuntary from intended sounds; the distorted sounds of dysarthria that compel harder, better listening; and the lost or garbled words of aphasia that require us to use gestures or draw pictures, opening up entire communication experiences we otherwise might never have. Viewed one way: these are small prices to pay for the linguistic inventiveness, deeper connection and unique insights that speech disorders also bring.

These, then, are some of the productive qualities of speech disorders. Although it is rarely acknowledged, they have a positive, as well as negative, impact on an individual's personality: the desire to achieve, a tendency towards compassion, a unique wisdom. They change the way that language is used, in a way that is creative and imaginative as well as limiting. And they disrupt the otherwise predictable flow of language and communication, allowing for new experiences to emerge.

Aneurin Bevan was one of many millions who experienced these productive and ebullient qualities. Rather than saying he achieved in spite of a speech disorder, we can see how it was because of it. His stutter gave him ambition, ensuring that he was a famous orator before he was even twenty, as well as informing the compassion that saw him relentlessly fight for and create the National Health Service. It caused him to studiously cultivate a wide vocabulary and range of synonyms, making him one of the most articulate, even poetic of parliamentary adversaries. And the act or moment of stuttering could disrupt the usual formalities and patterns of debate, allowing him to wield what he called the 'poniard of truth'.

These three tendencies – of personality, language and communication – are the bedrock for a neurodiverse appreciation of speech disorders. Those with such conditions have the potential to contribute to a more empathetic, linguistically innovative and communicative society, yet rarely get the chance because they are so discriminated against. There is a beauty and even a logic to these conditions that is smothered beneath the word 'disorder'.

I believe that as they become more accepted and integrated within society, we will all find ourselves learning to communicate in new and subtly different ways, for to engage with a person who stutters, tics, has aphasia or dysarthria on their own terms requires better listening, less interrupting and looking beyond the words people say to the other ways in which they may be communicating. It means everyone reaping the benefits I have gained from the hundreds of interactions over the past few years that have only enriched my understanding of human nature and language.

This is just the foundation. Around it develop other productive traits that might otherwise seem merely speculative, but follow as a consequence. If the combination of these qualities in an individual can stimulate enhanced and unique creativity in verbal communication, in the right hands speech disorders also inspire great art.

9

The Art of Disorder

As long as he could remember, Charles Dodgson struggled with speech. He might be halfway through a sentence and suddenly a word would stick in his throat and he'd be left mouthing silently like a fish out of water. At home, it didn't matter: most of his siblings were the same. They were a big household, tucked away in the country. A micro-community of their own. Charles was the family joker and wrote a poem gently mocking their strong-willed father, describing the 'Rules and Regulations' of the household. 'Learn well your grammar, and never stammer' it begins, followed by a list of petty instructions: 'love early rising', 'go walk of six miles', 'shut a door with a handle', then 'once more, don't stutter.'[1]

At school it was bad. The other boys were noisy and sporty and quick to spot weakness in others. Charles was a bookworm with a speech impediment. All the hardship and lack of friendship was summed up in the school-book where, underneath his name, someone wrote '…is a muff'.[2] That's all he was to them: a fool, a weakling, a stutterer. He got through his time, but left knowing that nothing on earth would make him go through such experiences again.

The day he arrived at Christ Church to study mathematics, he knew he never wanted to leave. Other people loved Oxford for the society, but for Charles it was the reverse. It was quiet and solitary. People left him alone. He was happy. Fellow students struggled to remember him: he seldom spoke and his impediment was not, as one contemporary recalled, 'conducive to conversation'.[3] After graduating, he simply joined the faculty, spending his days researching obscure mathematical problems. In his spare time, he developed a few hobbies: drawing, poetry, photography, writing letters back home. Things that took place away from human society. He became part of the furniture of the university; another eccentric loner. Over the years, though, he managed to make a few friends. 'Those stammering bouts were rather terrifying', remembered May Barber:

> It wasn't exactly a stammer, because there was no noise, he just opened his mouth. But there was a wait, a very nervous wait from everybody's point of view: it was very curious. He didn't always have it, but sometimes he did. When he was in the middle of telling a story... he'd suddenly stop and you wondered if you'd done anything wrong. Then you looked at him and you knew that you hadn't, it was alright. You got used to it after a bit.[4]

He drew a picture in a letter to another friend: 'a little thing to give you an idea of what I look like when I'm lecturing. The merest sketch, you will allow – yet still I think there's something grand in the expression of the brow and the action of the hand.'[5] There is nothing grand in the accompanying picture, though: it shows a man with a hand clamped over his mouth, eyes bulging in alarm, mute and ridiculous.

Occasionally, Charles dreamed of becoming a man of the world – a public figure, a great speaker. But he knew his stammer would let him down. He gradually accepted that a quiet life as a respected but uninspiring professor might be the best he could hope for. Then he heard reports of the speech therapist who was, by all accounts, a miracle worker. James Hunt had established a country retreat called Ore House where pupils could come and study together for weeks at a time. Charles Dodgson enlisted and spent the summer of 1859 participating in Hunt's pioneering experiment in group therapy.[6] Each day, clients were required to do certain activities together: mostly reading aloud, debating and delivering speeches. Often Hunt would leave them to it while he worked on the manuscript of *Stammering and Stuttering* in his study. In this safe environment, his pupils could address the often insurmountable fear of public speaking, or even just everyday conversation, and build confidence and self-reliance as talkers.

James Hunt's literary and philosophical approach to speech therapy would have appealed to Dodgson. Hunt was fond of quoting the philosopher John Locke's claim in *An Essay Concerning Human Understanding* (1690) that it is hard to determine whether 'language, as it has been employed, has contributed more to the improvement or hindrance of knowledge among mankind'. On the one hand, Hunt may have been emphasising the importance of his pupils mastering a better usage of words. On the other, he may have been encouraging them to be less tough on themselves. Rather than seeing language as perfect and their use of it imperfect, they should take courage from the fact that language itself is full of flaws.

It is the latter interpretation that seems to have influenced his pupils. For around Hunt gathered three writers, all with stutters, who were to prove pioneers of a new genre of literature: the modern fairy story, or what would become fantasy fiction. There was Charles Kingsley, who after being treated by Hunt wrote a public endorsement of *Stammering and Stuttering* and went on to write *The Water-Babies*. There was George MacDonald, future author of *The Princess and the Goblin*. These iconic books share a playful use of language, creating worlds where you can meet Mrs Doasyouwouldbedoneby or Prince Harelip, and visit strange places like Gwyntystorm. And, of course, there is the work of Charles Dodgson himself, that he would publish under the pseudonym Lewis Carroll.

Charles continued to visit James Hunt at Ore House in the years after 1859. During this time, he began experimenting with language in different ways. He wrote strange notes to friends that he called Puzzle Letters. Words were mixed with drawings and squiggles like a form of hieroglyphs; or they were reassembled into forms that require some decoding. There were mirror letters, back-to-front letters, circular pinwheel letters, fairy letters in tiny writing and letters with riddles and acrostics.

He also took ever greater delight in the company of his colleagues' children. He would invite them with their mothers or nannies to his rooms at Oxford, where he would take photographs of them in fancy dress. To entertain them, he would tell strange stories invented on the spot. Their questions and suggestions would send the narrative in unexpected directions. And all the time he would be preoccupied with taking his photographs. 'In this way,' remembered Alice Liddell, the little girl whose name became the title for one of the most famous books in the English language, 'the stories, slowly enunciated in his quiet voice with its curious stutter, were perfected.'[7]

One summer's afternoon in 1862, *Alice's Adventures in Wonderland* stuttered into existence. Dodgson composed it on the spot, verbally side-stepping, darting off in new digressions, much like the adaptive techniques he used to get through a conversation. He and a friend had taken Alice Liddell and her two sisters for a jaunt in the country. It was overwhelmingly hot and they took refuge in the shade of a hayrick in a meadow. As they rested, one of the sisters begged Charles to tell one of his whimsical stories. He began and couldn't stop. The children were entranced. His friend couldn't help interrupting, asking if it was improvised. 'Yes,' Charles said, laughing, 'I'm making it up as we go along.'[8] Later, when this moment had become legendary, Charles recalled how 'in a desperate attempt to strike out some new line of fairy-lore, I had sent my heroine straight down a rabbit-hole, to begin with, without the least idea what was to happen afterwards'.[9]

I believe *Alice's Adventures in Wonderland* and its sequel, *Through the Looking-Glass* (both published under his pen-name Lewis Carroll), are books that could only have been written by a person with a speech disorder.

Most writers use language to try and capture ideas, to convert abstract thought into argument and story. They value linguistic precision. Dodgson does the reverse. He loves language that is unruly and can't quite be controlled, as if he wants to continually remind us how unreliable it is, how little it can be trusted.

For instance, when Alice first arrives in Wonderland she cries so much at being lost in this strange place that she almost drowns herself and a gathering of animals in her own tears. Having pulled themselves onto land, the question arises of how to get dry, which prompts a 'very dry lecture' by a Mouse on William the Conqueror. This same Mouse then announces it will tell 'a long and sad tale' which transpires to be a concrete poem in the shape of its own tail. This is less a plot than a series of bad puns that rest on the potential of words to mean different things in different contexts. At other times, the story grinds to a halt entirely as when Alice's attempt at a logical discussion with the Red and White Queens in Looking-Glass land is rendered impossible because of their incessant word play. You can almost hear the voice of James Hunt behind it all: quoting John Locke, describing how unreliable and tricky language is, telling his pupils to be less uptight and more forgiving of themselves.

In his essay 'The Stuttering of Lewis Carroll', linguist Jacques de Keyser identifies Dodgson's use of portmanteau words as a consequence of his unusual speech.[10] Portmanteau words are created when two words are crushed into one, as in 'motor' and 'hotel' becoming 'motel'. They are accidentally created by children when stuttering or cluttering. Charles Dodgson's portmanteau creations include one-offs like 'borogove' and 'uffish', as well as a few which have joined the English language, like 'chortle' (chuckle/ snort) and 'galumphing' (gallop/triumphant).

One of the first things Alice does after stepping through the looking-glass is to pick up a book lying on a table and read 'Jabberwocky'. In gloriously arcane and invented language, the poem recounts a young hero's successful attempt to slay the dreaded Jabberwock: a mythical beast of Dodgson's own invention. 'The Anglo-Saxon word 'wocer' or 'wocor' signifies 'offspring' or 'fruit', Dodgson explained later: 'Taking "jabber" in its ordinary acceptation of "excited and voluble discussion", this would give the meaning of "the result of much excited and voluble discussion.""[11] The Jabberwock is, literally, the fruit of jabbering: also known as nonsense. But jabbering is symptomatic too of certain forms of stuttering and also cluttering, where speech becomes so fast and animated that it collapses into babble and repetitive sounds. Astonishingly, Dodgson turned such speech into one of the best-loved poems in the English language. De Keyser concludes: 'most probably Carroll used the phenomenon of stuttering, with which he was confronted daily, as a literary device, as a kind of positive contribution to the language in reaction to the negative influence of stuttering on his personality.'

Despite continuing to study with James Hunt, Charles Dodgson never overcame his stutter. In his last book, *Sylvie and Bruno*, published when he was sixty, he even appears as a stuttering narrator. And it was still foremost in his mind at the time of his death in 1898. In a letter composed just nine days before he died he wrote, 'the hesitation, from which I have suffered all my life, is always worse in reading (when I can see difficult words before they come) than in speaking. It is now many years since I ventured on reading in public – except now and then reading a lesson in College Chapel. Even that I find such a strain on the nerves that I seldom attempt it.'[12]

In the century since his death, people have searched for other appearances of Dodgson in his work. It has been suggested that the Dodo in *Alice's Adventures in Wonderland* represents a stuttered abbreviation of his own surname. More convincing is the notion that he appears as the White Knight at the end of *Through the Looking-Glass*. In this guise, he takes his leave of Alice Liddell, who was, by the time of publication, already grown into a young woman. The White Knight escorts Alice before her last move across the chessboard landscape of Looking-Glass land to become a queen, but he cannot make the move with her. Everything about this clumsy, eccentric knight represents a form of stuttering: he can't even stay on his horse, but falls off with a crash every time he gets back on. And when he sings his song, which has a range of different titles like 'Ways and Means' or 'Haddock's Eyes', he sings of an old man he used to know, whose look was mild, whose speech was slow, and muttered mumblingly and low, as if his mouth were full of dough: a fairly accurate description of Dodgson's 'hesitancy'. When at last he can accompany her no further, Alice races on ahead, pausing briefly to look back upon this kindly if somewhat ridiculous man – but only briefly, being far more preoccupied with the exciting future before her.

Lewis Carroll's speech disorder not only inspired him

to make art, but shaped the form and plots of the stories he wrote. To his name can be added a long list of artists with a speech disorder. I have already referred to writers, actors, scientists and philosophers like Elizabeth Bowen, Marilyn Monroe, Charles Darwin and Stephen Hawking, and there are many songwriters too, like Marc Almond, Edwyn Collins, Noel Gallagher, Kendrick Lamar, Carly Simon and Bill Withers. They are all artists who deal with words. On the page, on the stage, in song: words are easier to manipulate in these formal environments, away from the haphazard chatter, the give and take, of social interaction. 'The central irony of my life remains that my stutter, which at times caused so much suffering, is also responsible for my obsession with language,' writes American author Darcey Steinke. 'Without it I would not have been driven to write, to create rhythmic sentences easier to speak and to read. A fascination with words thrust me into a vocation that has kept me aflame with a desire to communicate.'[13]

In certain cases, art may be more than creative liberation but the only way of communicating beyond carers and institutions. Memoir writing by people with dysarthria or non-verbal conditions has become a literary genre in its own right, from Jean-Dominique Bauby's *The Diving Bell and the Butterfly* to Joey Deacon's *Tongue Tied* and Jonathan Bryan's *Eye Can Write*. 'I couldn't speak with my lips,' writes Christy Brown, author of *My Left Foot* and *Down All the Days*, describing the moment he discovered he could write with his toes. 'But now I would speak through something more lasting than spoken words – written words. That one letter, scrawled on the floor with a broken bit of yellow chalk gripped between my toes, was my road to a new world, my key to mental freedom. It was to provide a source of relaxation to the tense, taut thing that was I, which panted for expression behind a twisted mouth.'[14] These books are powerful works of art written by people for whom every word is chosen more carefully than those of even the greatest poets: they are compelling, sometimes strange reports from a world the rest of us cannot begin to imagine.

Alongside this long list of the famous are the many millions who have turned to art, not to make their name, but simply to facilitate the communication that is denied them by speech. I have always been most fulfilled when writing, making music or films. Most of all, I am drawn to people who do these things well, which is why I have always worked close to artists. I love seeing the connection between performers, creators, writers and their audience: forms of communication that go far beyond what is possible in daily conversation. Art fills me with the hope I never quite found in speech. Amy Samelson, a clinical social worker, captures this optimism when describing her work with people with aphasia: 'Creativity is in play, and this is where it gets interesting. Two people with good will and intention, wishing to reach one another, work to find ways to reach out. Sometimes single, words, drawings pictograms, written words (when possible) … and pantomime are part of the new language.'[15]

On 18 February 2005, the Scottish songwriter Edwyn Collins was interviewed by BBC Radio 6 Music about his new album. He had composed one of the indie anthems of the 1980s in 'Rip It Up' and had a worldwide hit with

'Girl Like You' in the 90s. There was some curiosity about his next release but, at forty-five, most fans supposed his best years were behind him. Famously outspoken, Collins was uncharacteristically lacklustre in the interview and explained he was feeling unwell: a nausea and vertigo he attributed to food poisoning. Back home, the symptoms worsened and he was soon hospitalised. In quick succession, Collins suffered two cerebral haemorrhages that put him into a coma for a week and saw him bedridden for the next six months. Complete recovery was impossible. The stroke had left him severely paralysed on the right side of his body and with acute aphasia, but he had youth on his side.

Over the following years, much of his speech returned, although he faltered and stuttered continuously. Aphasia continued to create a fog in his brain. Yet he found there were unexpected compensations: his creativity was not only unimpaired but enhanced. Although conversation was difficult, he experienced no problem with words when he sang and so songwriting became a vital way of communicating the innermost hopes and fears he struggled to express in speech. He had lost the use of his right hand, but he found a way of playing the guitar using his left alone. And he rediscovered a love of drawing. 'My brain relaxes when I draw, and can see the pages unfolding,' he told a journalist later. 'Before my stroke. My drawing. Was taking me ages to do it. For example a wigeon [a type of duck], but … one week to do! But after my stroke, I became free. As a bird! I came terribly relaxed, and I'm working on things, a quick sketch, and it's much better than the "artiste" stuff I did before the stroke – I became much freer.'[16]

In 2010, Collins surprised his fans and the music press with the release of a new album. *Losing Sleep* charted his emotional journey to acceptance and partial recovery. The lyrics, by his own description, had the 'simple language' of aphasia: 'Losing sleep, I'm losing sleep, I'm losing dignity, everything I know is right in front of me, and it's getting me down.' 'My new style? It's simple and direct,' Collins has said. 'But I like it. I used to be an intellectual; my words were complicated. Now, I'm straightforward. I have to be.'[17] The cover was a collage of his drawings of British birds.

Collins had depended upon the contributions of friends to the songwriting, but his next two albums were all his own.[18] The clever wordplay of his early songs had returned. These were not just recovery albums, but among the best work of his entire career. Collins has never entirely regained his previous capacity with language – he finds it hard to read or speak in fluent sentences – but as an artist his work has evolved and is revered by fans and critics. He has released records, toured the world, collaborated on a film about his experience and published a book of his illustrations. There has been a rare longevity and inventiveness to his career that arguably might not have been sustained if he had never experienced stroke and aphasia.

If speech disorders are a spur to art-making, they can also shape the form and style of output. After all, Edwyn Collins's work over the last decade is not what it would have been if he had never had aphasia. The word Aristotle used to describe the study of art (in his case, drama) was 'poetics'. Since then, the term has been used to describe a wide range of creative forms with their own internal rules and tendencies: from poetry to Caribbean patois, classical music to cinema. Is it possible, then, to talk about a poetics of disfluency, identifying qualities that run as enduring concerns through the art of people with speech disorders?

The story of Charles Dodgson/Lewis Carroll is important in this regard because he is the earliest artist of any discipline whose biography offers enough detail to draw specific links between his work and his speech disorder. He is also the first to have consciously done so, acutely aware of the strangeness of both his spoken and written use of language. But it is compelling too because the *Alice* books are among the most influential and loved in the entire literary canon. They are, therefore, a plausible and precious keystone in a neurodiverse account of speech disorders. This, the earliest work of art we can claim with some certainty to be influenced by a speech disorder, also contains the enduring, productive qualities that artists who stutter or have other speech disorders have shared since.

I will describe each in turn, but they include the unique insight (having 'something to say') that is key to making art and comes in part from the experience of disfluency: the way Dodgson often kept to himself or hid behind his camera in social situations, but let his words fly in bizarre and compelling directions when with children. It includes the linguistic dexterity and playfulness that those with speech disorders develop as a way of concealing or mitigating their condition but can have startling results when used in art-making. And it includes a tendency to experiment in form; to break the rules of language that hamper a disfluent speaker's ability to participate in conventional forms of communication.

As we have seen, there are many personality traits common to people with speech disorders. Many are negative: a sense of isolation and secretiveness, of carrying a stigma, sometimes of despair. But there are positive ones too: a drive to succeed against adversity, an acquired empathy and a unique wisdom about human experience. These are all qualities that are frequently associated with those of the artist. Since art depends on the originality and authenticity of 'voice' (in its broadest sense), and since in any case no maker can remove their personality from their output, these qualities may be ones that consistently translate into the creations of those with speech disorders.

In his book *Stutter*, cultural historian Marc Shell wonders if there could ever be a 'Stutter Culture'. If so, it 'would be informed by awareness of isolation as an inescapable condition'.[19] This is certainly true of Charles Dodgson, who was immensely shy in person and heard James Hunt's theories about the 'habits of secrecy' that emerge from stuttering directly from the man himself. Charles Dickens was a fluent speaker, but he nevertheless observed astutely that 'stammering rises as a barrier by which the sufferer feels that the world without is separated from the world within'.[20]

One person who managed to turn this psychological predicament into art was Somerset Maugham. Through a long and productive career, he emerged as one of the most famous writers of the twentieth century, best known for novels like *Of Human Bondage*, *The Moon and Sixpence* and *The Razor's Edge*. 'The first thing you should know,' he once said, 'is that my life and my production have been greatly influenced by my stammer.'[21] While he discusses his impediment in some autobiographical sketches, it isn't immediately clear how his 'production' has been shaped by it. After all, in his vast oeuvre of plays, novels and stories, there isn't a single character who stammers. The closest is the impediment he gave to his alter ego Philip Carey in the semi-autobiographical novel, *Of Human Bondage*: not a speech disorder though, but a club foot.

The influence of stuttering seems instead to lie in the mood or atmosphere of his work. Maugham was a loner: he hated talking to strangers because he stuttered so much when he did and depended on garrulous lovers to manage social intercourse on his behalf. In later life, he created a loner's paradise for himself in the secluded Villa La Maur-esque on a promontory of land jutting from the French Riviera into the Mediterranean Sea. The parties there were famous but Maugham, like Gatsby, would drift in and out according to his fancy. The narrators of Maugham's stories are generally loners like himself, whether it's the 'British Agent' Ashenden (Maugham himself worked as a spy for the British government at various times in his career), or the voluntary exiles eking out their existence in the furthest posts of the British Empire in works like 'Rain' and *The Painted Veil*. In *The Moon and Sixpence*, the narrator states:

> Each one of us is alone in the world. He is shut in a tower of brass, and can communicate with his fellows only by signs, and the signs have no common value, so

that their sense is vague and uncertain. We seek piti-
fully to convey to others the treasures of our heart,
but they have not the power to accept them, and so
we go lonely, side by side but not together, unable to
know our fellows and unknown by them.[22]

The loneliness that pervades Maugham's work is frequently attributed to his homosexuality, despite the fact
that Maugham was, comparatively speaking, one of the
most openly gay public figures of his day. It is a theory
that ignores what Maugham himself claimed: his stutter,
not his sexuality, was the 'first thing' one should know
about his writing.

Many other writers with stutters describe how a sense
of isolation and stigma cultivated at a young age finds
expression in their work. The Irish novelist Colm Tóibín,
author of *Brooklyn*, tells me his stutter arose after a traumatic episode when he was eight. His father had a brain
haemorrhage and he was sent away.

> I was staying with an aunt, and I didn't know where
> everyone else was. I had been told I was safe and every-
> thing was okay, but it wasn't and I didn't know where
> I was or when I was going home. I always knew that
> some damage had been done to me. That experience
> of what happened in those months never left me. And
> if I'm writing I find it very easy to evoke that period.
> I have written about it a good number of times. It's a
> hurt. And the stammer was a symptom of that.

Brian Catling, sculptor and writer, tells me his stutter

means he is 'fascinated by all abnormalities of behaviour. I have to be careful: I can find myself following odd people round supermarkets because I feel a kinship.' All of his output, including the recent *Vorrh* trilogy, focuses on human abnormality: characters who are deformed, unhinged and lonely.

Because of this sense of isolation, of being an outsider, stuttering has a special relationship with the musical genre most associated with those same qualities: the blues. Two of the greatest bluesmen of all made this connection explicitly. 'As a child, I stuttered,' B.B. King wrote in his autobiography. 'What was inside couldn't get out. I'm still not real fluent. If I were wrongfully accused of a crime, I'd have a tough time explaining my innocence. I'd stammer and stumble and choke up until the judge would throw me in jail. Words aren't my friends. Music is. Sounds, notes, rhythms. I talk through music.'[23]

John Lee Hooker had a similar experience and impersonates himself in 'Stuttering Blues', where the singer's attempts to seduce a woman are undermined by his speech. 'Excuse me, baby, I can't get my words out just like I want,' he sings, stuttering as he does so. Then, with perfect fluency, 'but I can get my loving like I want it'.[24] According to biographer Charles Shaar Murray, Hooker's first producer 'claims to this day that his primary reason for deciding to record the young bluesman in the first place was that he was intrigued by the notion of a man who stuttered when he spoke, but not when he sang'.[25]

A certain romantic view of stuttering echoes in the popular rock music that the blues inspired. This speech disorder, so stigmatised in real life, in the mouths of rock

and roll singers is charismatic and sexy. In 'My Generation', Roger Daltrey of The Who pretends to stutter in
every line without an obvious reason why (although he
puts it to great effect when he sings 'why don't you all …'
and then blocks on an 'f', raising an appalling possibility for radio broadcasters in the 1960s, before following
through with 'fade away').

In the early 1970s, when performing his song 'Cyprus
Avenue', Van Morrison, a disciple of John Lee Hooker,
would stutter for a long time on the word 'tongue-tied',
a moment that always drew catcalls and cheers.[26] 'Morrison's work has always in part been about words failing,'
critic Laura Barton recently wrote in the *Guardian,*
'about inarticulacy and the gulf between the emotion and
the tongue. Many of his songs seem to sit at the precise
point where language falls apart.'[27] On a live recording
of 'Whole Lotta Love' from 1972, Led Zeppelin's Robert
Plant fakes a long stutter on the final 'b' in the phrase
'You've got to let that boy boogie' before saying, 'I think
John Lee Hooker said that.'[28]

There are several musicians who have claimed that
singing and songwriting were a refuge from stuttering. 'As
a kid, I used to stutter,' Kendrick Lamar has said. 'I think
that's why I put my energy into making music. That's
how I get my thoughts out.'[29] Ed Sheeran claims learning
every word of Eminem's *The Marshall Mathers LP* when
he was ten made him fluent: 'he raps very fast and very
melodically, and very percussively, and it helped me get rid
of the stutter.'[30] Carly Simon creates a charming image,
straight out of *The Sound of Music*, of her family literally singing her to fluency. 'Because everyone in the family

knew that it was very hard for me to speak but very easy to sing,' she has said, 'we began to sing around the house all the time, telling each other to go to bed, or get up, or come to dinner.'[31] But while all these artists were drawn to music because of their speech, only the blues and its derivatives have made music about stuttering, turning it into an overt act of rebellion and seduction.

The unique insights and emotions that emerge from the experience of a speech disorder inevitably influence not only an artist's vision, but also their use of language. Again and again, writers who struggle with speech have described their elation at the way words flow on the page. Colm Tóibín tells me his stutter is reflected not in 'the style so much as the wanting to do it, as much as the finding comfort in it … One never really knew how problematic the speech was because you buried it. You put it aside. But when you had a pen in your hand you didn't have that problem at all. So there was almost pleasure but there was certainly release.' Poet and novelist Owen Sheers, who also stutters, describes to me the 'release and satisfaction in being able to be fluent on the page and to be very precise about words and their order'.

Unsurprisingly, the stuttering writer has often been prone to graphomania: a compulsive need to get words down because so many get caught in the mouth. Somerset Maugham wrote over thirty books of fiction, over thirty plays, and countless stories and volumes of essays and memoirs. Henry James wrote over twenty novels and over a hundred stories. Although Lewis Carroll published relatively little, he funnelled his creativity into his imaginative letters: over 100,000 in the course of his lifetime.

John Updike was scarcely less prolific and wrote about the satisfaction he took in having 'managed to manoeuvre several millions of words' around the 'guilty blockage in the throat'.[32]

Although the term Tourette's syndrome didn't exist in his lifetime, the eighteenth-century writer Samuel Johnson has been retrospectively diagnosed, based on the close observations of his friends Fanny Burney and James Boswell. He wrote many essays and biographies as well as a novel and a play, but his fame in his lifetime rested on his *A Dictionary of the English Language* (1755). At the time, and to this day, the task seems superhuman. There was no comprehensive English dictionary at the time; certainly nothing to compare with the French dictionary produced by the Académie française a half-century before. As Johnson himself famously observed, if it took forty scholars forty years to complete the French dictionary, he would do his in three. 'Let me see,' he joked, 'forty times forty is sixteen hundred. As three to sixteen hundred, so is the proportion of an Englishman to a Frenchman.'[33] In the preface to his dictionary, Johnson repeatedly talks about wanting to 'fix' the English language: not to correct it so much as pin it down so that words 'might be less apt to decay and that signs might be permanent, like the things which they denote'.[34] He sees language itself as unreliable, like the tics, gesticulations and involuntary ejaculations others observed in him.

More than the sheer volume of output, people with speech disorders often develop a unique style of writing that reflects some of the patterns of their speech. 'The way I write is very different to the way I talk,' says actor and director Jamie Beddard, reflecting on his experience of cerebral palsy. 'The language I use in speech is very functional because I know what works and what you're not going to understand. But on the page, I'm very pompous because I can't be pompous in real life.'

Earlier, we saw how Henry James's strange, circumlocutious way of speaking was the result of an interiorised stutter. In the autumn of 1896, a wrist condition prevented him from writing by hand and he was forced to employ a stenographer. Overnight, he went from writing novels to dictating them. As biographer Leon Edel observed, 'Henry James writing, and Henry James dictating, were different persons. Some of his friends claimed they could put their finger on the exact chapter in *Maisie* [*What Maisie Knew*] where manual effort ceased and dictation began. After several years of consistent dictating, the "later manner" of Henry James emerged.'[35]

The novels in the 'later manner' include *The Turn of the Screw*, *The Wings of the Dove*, *The Ambassadors* and *The Golden Bowl*. In them, he abandons the crystal clarity of his early work for a style that is obscure to say the least: long, very long sentences and sophisticated grammatical structures, or what has been called the 'complex' and 'indirect' style of these books. Characters no longer say what they mean; in fact, they tend to do the exact opposite. In *The Golden Bowl*, we are witness to a series of scenes charting an adulterous affair in which everything occurs beneath the surface of dialogue which is either so shrouded in secret hints and signals it is almost impossible to make sense of it, or completely off target altogether concealing some deeper form of communication. James's point is that human behaviour is too complex, too con-flicted, too profound to reduce to definitive statements and articulate dialogue.

Most artists innovate in the early stage of their career, but James did so at the end, leaving behind the epic, mor-alising sweep of Victorian masters like George Eliot and Charles Dickens for something simultaneously vaguer but celebratory of human nature as it really is. This small group of books is recognised as a high point in the history of the novel, both because of their quality but also for the widespread influence they had on other novelists, funda-mentally changing the art form for good. While there are many things that informed James's late, great period, the shift from writing to dictating is widely acknowledged as an important one. What we see is the carefully consid-ered, circumlocutory, vast vocabulary and grammatical dexterity of the stutterer transformed with Midas-like alchemy into art.

Aphasia is known for depriving people of language and yet it also has its own poetic turn. After his stroke, Edwyn Collins could only say four things. Three of them were functional: 'yes', 'no' and 'Grace Maxwell' (his wife's name). But the fourth was seemingly random: 'The possibilities are endless.' Today, Collins has no idea where this phrase came from or what was really meant by it, except possibly being an enigmatic reflection on his own potential as a recovering stroke victim or a state-ment about life itself. As his recovery progressed, other strange phrases emerged from his mouth. His wife kept a record of them, noting that he wasn't 'really able to control these surprising announcements'. They included:

'Subtle differences' – 'The situation is evolving' – 'Life is an aphrodisiac' – 'Suffering is ordinary. Suffering is the place he is understanding. Means towards an end' – 'Different voices come. It rocks my world' – 'I think to myself, is it any wonder he's gone mad?'[36]

One writer who became fascinated by the seemingly random, yet poetic language of aphasia was Irish playwright Samuel Beckett. As a young man, he found himself questioning the 'official English' he heard and read all about him. It was 'like a veil that must be torn apart in order to get at the things (or the Nothingness) behind it,' he wrote to a friend. 'Grammar and style. To me they have become as irrelevant as a Victorian bathing suit or the imperturbability of a true gentleman. A mask. Let us hope the time will come, thank God that in certain circles it has already come, when language is most efficiently used where it is being most efficiently misused.'[37] For Beckett, speech disorders were often exemplary examples of efficient misuse. In the 1930s, he visited the French poet Valéry Larbaud who, following a stroke, could only speak a single sentence: 'Bonsoir, les choses d'ici-bas' (sometimes translated as 'Farewell material things of the earth').[38] Literary scholars have repeatedly pointed to the similarities between Beckett's theatrical masterpieces, such as Not I, and the transcribed speech of certain forms of aphasia. At the end of his life, Beckett experienced aphasia himself and his last poem describes the experience of trying and failing to articulate himself with the constant refrain: 'what is the word?'[39]

While a speech disorder can make an individual revere conventional artistic form, to excel on the page or on stage where they cannot in day-to-day speech, it can also push them the other way, to experiment. In some cases it is a natural consequence of that non-linear, spontaneous, almost three-dimensional way of thinking that accompanies a disorder. This is evident in the work of Charles Dodgson, who claimed he had no idea what he was up to when he composed the *Alice* books, and Henry James, whose late style effectively redefined the rules of the novel.

The American writer David Shields is mid-way through a career intent on taking this tendency to a logical conclusion. When he was in his twenties, he achieved literary success with his novel *Dead Languages*, a semi-autobiographical story about a boy growing up with a stutter, as Shields had done. *Dead Languages* is a conventional coming-of-age novel in many ways, and the perfect expression of his youthful desire to turn the nightmare of 'stuttering into lyric language'. 'Writing *Dead Languages* freed me up on some unconscious level to take stuttering no longer as subject matter nor as default mode of lyrical writing,' he tells me, 'and pivot into making stuttering the very methodology of my new found art form, namely literary collage, or montage or fragmentation.'

Shields began, tentatively at first but with ever greater ferocity, writing books which were neither fiction nor non-fiction but merged different genres, that were non-linear and often fragmented without connecting thoughts between sections, and that were built around the principle of collage, often putting in extracts from other writers' work. He has described this form as the 'lyric essay'. His influential 2010 manifesto *Reality Hunger* argues that conventional narrative form, or the novel, is no longer adequate for capturing the experience of modern life. Only work which merges different forms, shuttles between fact and fiction, leaves out long-winded set-ups and bridging thoughts, and in which the authorship of any single passage is unclear (all of which, incidentally, are qualities we associate with the internet) can do this. 'I am terribly distrustful of fluency of all kinds,' Shields tells me. '*Reality Hunger* is a sort of stutter-fest in a way. Stuttering understood as a metaphor for discordant literary form.'

In his article 'Tourette's Syndrome and Creativity' (1992), Oliver Sacks describes a writer who separates his practice: short, formal essays when he is repressing his Tourette's and 'huge, meandering, fantastical (and often coprolalic) novels, in which he gives his Tourettic fancies full reign'.[40] This description applies well to the theatre work of Jess Thom. Her show *Backstage in Biscuit Land* offered a unique insight into the surreal comedy as well as the difficulties of living with her condition. Rather than trying to write her tics out of the script, Thom embraced them. This included those that emerged during the production process. When she was asked what props she wanted for the show, she let her tics fly and, as one reviewer described it, on the stage were 'four ducks dressed as pterodactyls; a dinosaur balloon; an enormous loaf of bread named Steve; an anvil with the word "dinner" on it; a life-size statue of Mother Teresa; and "the smell of an ice cream parlour and bakery from a different age".'[41] The last one sprayed on the front row of the audience.

The British performance artist Brian Catling grew up with a stutter, dyslexia and tics, or what he has called 'the full set'. Although he suffered at school, it has been a long time since someone has dared mock him because, even in his seventies, he is an imposing presence: a highbrow bruiser like Dr Johnson. He began his artistic career as a sculptor, but as much as he loved making things with his hands it wasn't enough. He started writing poetry. Even then, there was something missing. Finally, for an event at the Whitechapel Gallery, he embarked, fairly spontaneously, on a work of performance art. He sat in the room tearing apart books and putting them together in new combinations.

Catling has created many performances since, but they tend to revolve around creatures played by himself, often for days at a time, with a heightened abnormality. This includes a cyclops (using a prosthetic mask) or a man with rape alarms attached to his head. 'All of the performance personalities are not heroes – the cyclops being the most obvious – all of them have an impairment that makes them more human,' Catling tells me when we meet at the Royal Academy in London. 'They're not there to demonstrate control of the world. They're not there to demonstrate a heroic posture or a posture of control. They're stumbling in it and so they become inventive.' Catling roots all of this back to his stutter: that feeling of struggling through the world with a handicap. 'I've consciously looked at verbal disability. I've made performance pieces about not being able to speak.'

Late in his career, Catling sat down one morning and began writing a piece of fiction that I think Lewis Carroll would have approved of. Not only was it a work of fantasy, but, like *Alice,* it began with a single image with no idea what would come next. He thought it would be completed in a day or two, but finished several years later. The *Vorrh* trilogy (*The Vorrh*, 2012; *The Erstwhile*, 2017; and *The Cloven*, 2018) became an immediate cult classic, celebrated as one of the great visionary works of the twenty-first century by Phillip Pullman, Terry Gilliam, Tom Waits and many others. Restlessly, Catling moves on to the next project: a sculpture show, a collection of short stories, a play and a film.

In the last chapter, we saw how a neurodiverse approach to speech disorders would recognise the immense difficulties people may experience with speech, but also acknowledge some of the productive qualities that may develop as well: the impact on an individual's personality, on their use of language, but also on the forms of social interaction they participate in. The unique forms of creativity that emerge alongside or because of a speech disorder mirror these areas shaping the personality or attitude of an artist, their use of language and imagery, and the tendency to experiment as they search for better ways of representing these insights. In other words, a speech disorder can impact on what an artist has to say and how they say it.

No wonder that Charles Dodgson had such a conflicted relationship with his own stutter. On one hand, he participated in intensive therapy for decades to try and alleviate it. But here is the strange thing: when he chose a pen-name for himself, a name he could choose out of any in the world, he settled on one that began with his most feared sound – a hard 'C'. This sound was so troublesome that Dodgson had to sometimes spell out what he wanted to say and even referred to it as 'my vanquisher in

single-hand combat'.[42] Every time he said 'Lewis Carroll'
he was reminded of his disfluency and the possibility it
might trip him up at that moment. Since it is hardly likely
this was a surprise to him, the likelihood is that he did
it deliberately: a constant reminder of both his greatest
shame and the oddity that made him unique.

10

Speech Acts of Resistance

Around 1916, a new front opened up in the Great War. It wasn't a territorial front, like the trenches surrounding Ypres or Gallipoli, for it was both invisible and pervasive, transcending borders, social divides and military loyalties. It was a conflict that had been simmering for many years but seemed suddenly to be bursting forth in the most unlikely of places. The target was immaterial: as light as air, as lethal as the biggest guns. Many of those participating in the assault were often unaware of their part in the conflict, but others had no doubt at all about their aim. This was a war against language.

Over the previous hundred and fifty years, since the emergence of what was called the Enlightenment in Europe, language had been increasingly claimed as the essence of human achievement: the vehicle of reason, of human co-operation and order. Countless politicians, philosophers, doctors, lawyers and clerics had repeatedly, and with ever greater force, expressed the view that our ability to speak and write was what separated us from other animals. Yet somehow this same civilising spirit, in the mouths and pens of those considered most adept at using it, had cajoled all of Europe into murder. The assassination of Archduke Franz Ferdinand in Sarajevo was incidental. After all, what did the murder of the heir to the Austro-Hungarian Empire by his own dissidents have to do with Britain? But for years, the words of priests, politicians, generals and polemicists had inveigled the paradoxical notion that war could prove a civilising force in the hearts of those they served. Within weeks of Franz Ferdinand's death, most of Europe was mobilising without most citizens quite knowing why, but nevertheless believing the voices of authority that insisted it was both necessary and noble.

The astonishing thing isn't that Europe went to war, but that so few resisted it. Doing so involved overturning every authority and certainty one knew. One of those coming to terms with this was the poet and soldier Siegfried Sassoon. Like millions, he had enthusiastically enlisted in the first months of the war. He was doing not only what the British Establishment wanted him to do, but what his friends, family, school masters and neighbours expected of him.

As an officer in the Welsh Fusiliers, he was revered for his bravery. Solo dashes into no man's land, even into enemy trenches, earned him a Military Cross and the affectionate nickname 'Mad Jack' from his men. To all appearances he was a fierce patriot, but the poems he wrote in private told a different story. While he had once written sentimental accounts of the honesty and loyalty of common soldiers, he increasingly satirised with terrible bitterness the authority figures who had betrayed them: a bishop who witters on about 'just cause' and 'the ways of God' when confronted by those whose lives have been wrecked,[1] and a general who is full of pleasantries with his men even as he devises the strategies that will kill them.[2] Sassoon later wrote that he wanted his poems to reveal what he called 'war's demented language'.[3] His friend Robert Graves described how British civilians were taken over by a 'war-madness' that appeared as 'a foreign language; and it was newspaper language'.[4] Both men still struggled to articulate it, but they sensed that the greatest enemy wasn't other humans but something rotten in the language we use.

In Zurich, Switzerland, this conviction was more flamboyantly expressed by a group of refugee artists and political activists, mostly from Germany, Romania and France, who were determined to sit out the slaughter in one of the few countries that had remained neutral. They declared themselves not only against reason and art, but against language too – or, at least, all existing forms of it – for it was words and argument that had created the war. They chose the nonsense word *dada* to describe their movement. On 14 July 1916, two weeks after the outbreak of the Battle of the Somme, the German poet and deserter Hugo Ball read out the first Dada manifesto in a public lecture. He described how language had become 'accursed' and covered in 'filth'. 'How can one get rid of everything that smacks of journalism, worms, everything nice and right, blinkered, moralistic, europeanised, enervated? By saying dada.'[5] He then illustrated his point with a shift into (semi) nonsense. 'Dada is the world soul, dada is the pawnshop. Dada is the world's best lily-milk soap. Dada Mr Rubiner, dada Mr Korrodi. Dada Mr Anastasius Lilienstein.' The poems he wrote 'meant to dispense

with conventional language and to have done with it.'
Like the name of the movement, they were nonsensical,
full of made-up words and sounds.

The spirit of Dada wasn't isolated to a group of oddball
deserters holed up in a mountain town. In Russia, a crew
of poets was experimenting with an art form they called
Zaum. One of them, Aleksei Kruchenykh, argued that
language binds us to dominant hierarchies and ideologies.
His manifestos call for a 'language that does not have any
definite meaning (not frozen), a transrational language'.[6]
In France, the poet Guillaume Apollinaire experimented
in the trenches with what he called 'calligrammes': poems
where words are laid out pictorially on a page without
any sentence structure. Versions of the same iconoclastic
turn against language took place among the soldier-poets
of Vorticism in England and Futurism in Italy, although
these artists tended to celebrate war as the force that
would set language free once and for all.

All too often, avant-garde art movements sit in isolation, even against the broader trends of society, but
for once the artists and armies were in alignment. The
authorities weren't concerned about a few dissidents,
but they were by the thousands returning from the front
incapacitated by shell shock. Doctors, generals and politicians were at first reluctant to admit the cause might be
psychological. The symptoms, they argued, were physiological. It was a form of concussion caused by internal
damage to the nervous system brought on by the noise
and shudder of shell explosions: hence its name. As the
war progressed, the number of cases of shell shock developed into an epidemic. In Britain alone, it is estimated

that somewhere between 80,000 and 200,000 soldiers were ultimately discharged from active service for it.[7]

Although shell shock had a wide range of symptoms, including paralysis, trembling, anxiety and nightmares, it was the disorders of speech in particular that captured the public's imagination. These ranged from mild stuttering to complete mutism. One of the best-loved songs of the war was 'K-K-K-Katy' – 'The Sensational Stammering Song Success Sung by the Soldiers and Sailors'.[8] In his poem 'Survivors', Siegfried Sassoon identifies shell shock victims by their 'stammering, disconnected talk'.[9] His friend Wilfred Owen, while convalescing, wondered in verse if he should 'mutter and stutter and wangle my ticket' out of the front line.[10] He was conscious that his speech impediment, involuntary as it was, might also save his life; what in peacetime was stigma, in war was a ticket to survival. In Virginia Woolf's *Mrs Dalloway*, the stutter of its hero, Septimus Smith, betrays the lingering shell shock that will ultimately lead to his suicide.

I don't believe that the speech disorders associated with shell shock were a conscious reaction against the authoritative and supposedly rational language that had both brought about and was perpetuating the war. At the same time, the notion that some unconscious and collective resistance was at work cannot be easily dismissed. Today, shell shock is considered a misnomer for what we now call post-traumatic stress disorder (PTSD). However, it is important not to conflate the two terms. All forms of PTSD, argues historian Peter Leese, are 'culturally shaped' at any given time. Shell shock, therefore, was the unique expression of PTSD in the context and culture of early twentieth-century Europe. It was, Leese writes, a 'malleable, subjective state' that 'absorbs too the sympathy of comrade and relative, the outrage of editor and MP, the censure of officer and pension doctor'.[11] Somehow, in this context, stuttering, tics and mutism emerged as dominant symptoms, the only occasion in history when speech disorders have reached epidemic proportions.

In a recent study of over 300,000 veterans of the Iraq and Afghanistan wars, it was found that only 235 (0.08 per cent) had 'acquired stutters'.[12] Even allowing for the higher number of Britons who fought in the First World War, this still suggests only a few thousand with acquired stutters, if the symptoms of modern PTSD and shell shock were the same. While there are no such studies from the First World War, the regularity of speech disorders in doctors' case notes – 'complete loss of speech'; 'mutism'; 'no stammer previous to shock'; 'speaks in a halting fashion'; 'a hesitation in his speech'; 'a tremulous tongue' – suggests it was, as the cultural stereotypes imply, far more prevalent. Something about the context of the First World War made disordered speech a fitting response to the horrors that soldiers experienced. Peter Leese concludes that shell shock was, in part, 'a bodily collapse of reason and language', like that described by the Dadaist artists, but one that 'demonstrated more eloquently than any artistic statement the modern era's failure of words'. That language was somehow to blame seems to have infiltrated the experience of shell shock in a way never quite matched in wars since.

Attempts to treat shell shock were mostly dire, the aim being simply to get people back to the front as soon as possible and avoid it becoming a shield for cowardice. Attempts ranged from massage to electric shock therapy to, most brutally, what was called 'firing squad therapy', threatening capital punishment for cowardice and in some cases following through. For a while, the army actually pursued a stiff-upper-lip policy of mixing shell-shocked soldiers with 'cheery chaps' who had minor physical wounds (this prompted an organised response from one group of shell-shocked patients that they would rather risk another air raid than be submitted to the enthusiasm of another 'cheery chap'). During the Battle of the Somme, by autumn 1916, the army was reporting over 15,000 cases of shell shock a month.

It was against this backdrop, the complete failure of traditional medicine to explain or treat shell shock, that the previously fringe theories of Sigmund Freud, Sándor Ferenczi and other psychoanalysts began to gain credence in the medical establishment. One exponent was W.H.R. Rivers, who set up a pioneering practice at Craiglockhart Military Hospital in 1916. He drew in part from psychoanalysis, but differed in one important regard. While the workings of an unconscious mind clearly seemed at work in shell shock, with symptoms an expression of the anxiety soldiers struggled to articulate in words, Rivers thought it had nothing to do with the childhood sexuality that was a cornerstone of Freud's theories.[13]

Rivers refused to stigmatise shell shock and its accompanying disorders of speech, perhaps for no better reason than he had stuttered all his life. By strange coincidence, he was a nephew of James Hunt and had grown up around his uncle's practice, meeting devoted pupils like Charles

Dodgson. Rivers's shell shock treatment owed just as much to James Hunt's methods as it did to psychoanalysis. He accepted patients for who they were, without passing judgement. Although it's hard to know exactly what Rivers spoke to his patients about, for that took place in confidence between them, Siegfried Sassoon gives one revealing insight about his own sessions with him. 'We talked a lot about European politicians,' he recalled, 'and what they were saying.' In other words, they talked about language.

The Great War was a turning point; the moment when many, rather than a few, began to challenge the infallibility of language. As a consequence, it was also a moment when the perceived inferiority of speech disorders was briefly brought into question. Stuttering became so common as to hardly merit comment. When Sassoon mentions in passing that he sat for dinner at Craiglockhart between 'two bad stammerers',[14] one feels he could be talking about brunettes or bald men. The song 'K-K-K-Katy' is unlike almost any other cultural depiction of stuttering because it is sincere and romantic. Its hero, Jimmy, may stutter, but he is 'brave and bold' and loved by a beautiful woman. Rivers shrugged off any concern about stuttering, claiming it was nothing more than 'a defect of the brain, which gives contradictory orders simultaneously when disturbed in a certain way', and that the best thing was to 'forget it'.[15] A hundred years later, his theory increasingly seems correct.

In the same spirit, people with aphasia were no longer perceived as 'intelligent dogs', as one pre-war doctor described a patient,[16] but as ordinary human beings struggling with a disconnect between thought and words. The neurologist Henry Head, a close friend of Rivers, tended soldiers recovering from head injuries. His influential 'Aphasia and Kindred Disorders of Speech' (1920) not only advanced understanding of the condition, but finally recognised that most sufferers were no less intelligent than they had been before and described them with great dignity.[17] Unfortunately, it was a moment that was soon betrayed by the theoretical indulgences of psychoanalysis.

I've shown how speech disorders can exert productive, as well as negative, influences on an individual's personality, their use of language and their creativity, but this account of the First World War shows how they can have a broader social purpose too. In the right context, speech disorders can prove a necessary disruptor: challenging groupthink and literally blocking the flow of a society that has run amok with platitudes and empty rhetoric. In the 1970s, the French philosophers Gilles Deleuze and Felix Guattari articulated this in explicit terms. Their collaboration as writers and political activists followed the violent and, ultimately, failed uprising of May 1968, when French students and workers briefly took control of the streets.

In their sprawling masterpiece *Capitalism and Schizophrenia* (1972–80), Deleuze and Guattari set out to show how every aspect of Western society is focused on controlling and cajoling us into people we don't want to be.[18] This is achieved not so much through guns or tanks, although they play a part, but through language. Words are used to fix and limit things with names and labels: male/female, black/white, straight/gay. Language is always in cahoots with the inherently repressive 'megamachine' of the state to keep us in our place. It is what determines class and gender, what controls our thinking, what holds back our development as individuals.

Language, Deleuze and Guattari write, is made primarily 'to be obeyed and to compel obedience', it is 'generalised slavery'. Mess with language, their argument goes, and you start to mess with the very foundations of human oppression. They even suggest some of the ways in which this might be done, including 'indirect discourse', 'atypical expression' and enforced 'breaks and ruptures' in our speech. All these are qualities we associate either with disordered speech or the compensating tactics used to conceal it.

The term they use to describe this strategy is 'creative stuttering'; creative because it is voluntary and planned rather than compulsive. 'Creative stuttering', they say, is 'an affective and intensive language' or 'a poetic operation' that can make us pause and become more aware of the way our tropes and phrases limit rather than expand thought. 'Creative stuttering is what makes language grow from the middle, like grass,' they write, 'what puts language in perpetual disequilibrium.' Unsurprisingly, Deleuze and Guattari revered Lewis Carroll, who had both a 'creative' and an actual stutter, and in whose hands language becomes gloriously irrational and unreliable.

When I first encountered the work of Deleuze and Guattari, I found this phrase pleasantly shocking. I had never seen the word 'creative' placed before stuttering and it made me think about it in a new way. I had always seen my stutter as an embarrassment that shouldn't be allowed to disrupt conversations. Now I began to wonder if there wasn't a certain power to it. Later, when I began to practise voluntary stuttering through City Lit, I put this to the test: deliberately blocking on words in professional meetings while maintaining eye contact, seeing the momentary unease it gave people and using that to then drive home a particular point I wanted to make. But could stuttering or any speech disorder ever be revolutionary, as Deleuze and Guattari suggested? I began to search for precedents and eventually found one in the Zimbabwean writer Dambudzo Marechera.

Growing up under British rule in the 1960s, in what was then called Rhodesia, Marechera saw the role that language played in enforcing colonialism in his country. At home and on the streets, Marechera spoke his native Shona, but at school – like Ngũgĩ wa Thiong'o in Kenya – he had to speak English. In his mind, like thousands of others, he associated his own tongue with the chaos of the ghetto and that of his colonial masters with order and wealth. 'Shona was part of the ghetto daemon I was trying to escape,' he recalled.

Shona had been placed within the context of a degraded, mind-wrenching experience from which apparently the only escape was into the English language and education. The English language was automatically connected with the plush and seeming splendour of the white side of town. As far as expressing the creative turmoil within my head was concerned, I took to the English language as a duck takes to water. I was therefore a keen accomplice and student in my own mental colonisation.[19]

Whether it was as a consequence of that inner turmoil, or of the complexity of having to think and speak in two languages, or just coincidence, Marechera's speech failed him. 'I began to stammer horribly,' he said.

> It was terrible. Even speech, language, was deserting me. I stammered hideously for three years. Agony. You know in class the teacher asks something, my hand shoots up, I stand, everyone is looking, I just stammer away, stuttering, nobody understands, the answer is locked inside me. Finally the teacher in pity asks me to please sit down. I was learning to distrust language, a distrust necessary for a writer, especially one writing in a foreign language.[20]

Marechera got a place at the University of Rhodesia, but was expelled for his involvement in anti-colonial activism. His academic performance remained astonishing and he was accepted into New College, Oxford, but was soon expelled once more: this time for trying to set the college on fire. For the next few years, he was mostly homeless, sofa-surfing around the UK. And he started writing. If he couldn't burn down the buildings of the Establishment, maybe he could burn its language down. Could he write a book that would take all the confidence, that surety, of English and mangle it into something as uncertain and vulnerable as his native Shona had been rendered? A big ambition, but how to go about it? He found himself thinking back to that moment when he had begun stuttering:

There was the unease, the shock of being suddenly struck by stuttering, of being deserted by the very medium I was to use in all my art. This perhaps is in the undergrowth of my experimental use of English, standing it on its head, brutalising it into a more malleable shape for my own purposes. This may mean discarding grammar, throwing syntax out, subverting images from within, beating the drum and cymbals of rhythm, developing torture chambers of irony and sarcasm, gas ovens of limitless black resonance. For me this is the impossible, the exciting, the voluptuous blackening image that commits me totally to writing.[21]

In 1978, Marechera's *The House of Hunger* was published in the Heinemann African Writers series. The book describes his experiences growing up in a township in Rhodesia and the student protests he was involved in, but it is no ordinary memoir. The language is violent, often surreal. He writes about losing his speech, beginning to 'ramble, incoherently, in a disconnected manner' as he tries to reconcile two languages in his brain. 'When I talked it was in the form of an interminable argument, one side of which was always expressed in English and the other side always in Shona. At the same time I would be aware of myself as something indistinct but separate from both cultures.' In one passage, he dreams he is the boy we briefly encountered earlier, operated upon by the over-zealous Prussian surgeon Johann Friedrich Dieffenbach, 'snipping off chunks from the tips and sides' of his tongue. He is woken by his mother telling him his father has been run over and killed.

The plot of *The House of Hunger* is non-linear: a series of passages inspired by the fragmented speech of his stutter, often ending abruptly mid-sentence and jumping backwards and forwards in time. In this way, the whole book is an attempt to turn his speech disfluency into both creative writing and political activism. Running through *The House of Hunger* is an overwhelming anger at colonialism, a self-loathing for writing in English (the language which, whether he liked it or not, he had been taught to understand best) and a desire to mangle it. The book caused an immediate stir. Doris Lessing said it was 'like overhearing a scream'.[22] It won the *Guardian* Fiction Prize. At the awards ceremony, Marechera threw plates at the other guests. Tragically, he never managed to reconcile the two voices in his head but increasingly suffered from poor mental health (either a form of schizophrenia or manic depression). He returned to Zimbabwe when it became independent in 1982, but lived a homeless existence before dying of AIDS-related illness at the age of thirty-one.

Marechera used his stutter as a form of resistance to colonialism. Today, nearly forty years later, a young Swedish woman is using her difficulties with speech and communication to fight a different battle: against human extinction. Greta Thunberg emerged to global consciousness in the summer of 2018 for her part in launching Fridays for Future, a school strike movement demanding environmental policies from government to match the scale of the climate crisis. Images of her standing with painted slogans outside the Swedish parliament went viral. Within a year, in what has been called 'the Greta

Effect', she became the face and voice of environmental activism. Publishers directly credit her influence on a surge in children's books looking at the climate crisis.[23]

Undoubtedly, there is a power in the image of a vulnerable fifteen-year-old girl standing like David before the Goliaths of the world. But it is far more than that. When Thunberg speaks, she does so with an extraordinary bluntness and simplicity, repeating the same messages again and again. Most famously, she berated the leaders of the world at the 2019 United Nations Climate Action Summit in September of that year, accusing them of depriving her generation of a future through wilful inaction. 'How dare you!' she said again and again. In a talk earlier in the year, she tried to explain this aspect of her personality. 'I was diagnosed with Asperger's syndrome, OCD and selective mutism,' she said. 'That basically means I only speak when I think it's necessary. Now is one of those moments.'[24]

Far more than the Instagram images, I think Thunberg's power lies in her use of speech. She does not use any of the normal tricks and techniques of hyper-fluency because she can't. There is no humour, no attempt to rein in her emotions and charm or seduce her audience. She speaks directly, honestly and with unrestrained anger. The value of her words is only enhanced by the fact she has had to overcome the constraint of selective mutism to say them. When she says that she only speaks when it is necessary, it is obvious she's telling the truth. And because of that inner struggle, because of the unique way the words do come out, they have a currency greater than those of even the most revered orator. They are perhaps the best hope we have of forcing our governments to change course.

Speech disorders, whether intentionally or not, can become agents of resistance or conscientious objection to oppressive regimes or in times of war. Just as importantly, although a little less dramatically, they can also undermine the patterns of thinking that support such regimes and conflicts. When I was researching neurodiversity and speech activism, I spoke to Joshua St Pierre, a Canadian speech activist and founder of *Did I Stutter?*. But as an academic at the University of Alberta, his work also looks at fluency and disfluency in the context of philosophy. 'A goal for a lot of political philosophy is mutual understanding or consensus,' he tells me. 'But I'm critical of consensus as a goal because I fear that smooths over differences in the appeal to some kind of common space or common ground.' This reminds me again of the works of Plato which depict Socrates and his friends uncovering the tenets of a perfect society through a process of consensus building: asking questions, disagreeing, adapting statements, moving forwards. 'It always assumes a certain kind of speaker,' says St Pierre. 'A political actor who's able to articulate themselves and able to do so without any help.'

Studying such works, St Pierre was acutely aware that, in most people's eyes, his stuttered speech hampers him from fully participating in such debate. This is not self-censorship or paranoia. When classicist Christian Laes surveyed all existing references to speech disorders in the ancient world, he was struck by the fact they all referred to the barrier they were perceived to present to public discourse.[25] Since our democratic traditions have their roots in that time, disordered speech remains problematic for political and philosophical practice today. This is something politician Ed Balls discovered when he was accused of lack of confidence and poor debating style in the House of Commons simply because he stutters.[26] 'Sometimes my stammer gets the better of me in the first minute or two when I speak,' he has said, 'especially when I've got the prime minister, the chancellor and 300 Conservative MPs yelling at me at the tops of their voices.'[27]

For St Pierre, this inability to accommodate different styles or registers of speech suggests something flawed in the practice of political philosophy. 'I worry about things that are far too smooth and things that desire to be smooth,' he says. Eloquence and flow are at the heart of consensus building, all of which tends to obscure the flaws in a particular way of thinking. Speech disorders, on the other hand, disrupt that process. If one of Socrates's friends had a speech disorder, for instance, the dialogues of Plato would read very differently. Rather than jumping from statement to statement, arguments might digress or even just pause, allowing time for doubt about a particular, and even erroneous, train of thought to settle. And what does the widespread presence of speech disorders throughout the population say about a mode of hyperfluent reasoning that is supposed to be enlightened, but is nevertheless also discriminatory? 'To bring stuttering into the discussion,' St Pierre says, 'is to ask certain normative questions like why do we have to speak in certain ways to be taken up as rational, essential and political? Why does our world communicate in this highly technocratic and frantic way?'

Ludwig Wittgenstein, one of the greatest thinkers of the twentieth century, is somebody who did bring stuttering into philosophy.[28] As a young man, he came to believe the whole canon of western philosophy was built upon speculative and misleading wordplay. Language is more than flawed: it is a hindrance, preventing us from understanding the mysteries of existence. It leads us down mental culs-de-sac, wasting time on pointless exercises such as looking for the substance behind abstract concepts (for example, the question 'What is time?' assumes there is such a thing as 'time' to be defined). Any philosophy, Wittgenstein said, which is constructed around linguistic statements is nothing more than 'tidying up a room':[29] the mess being language itself; the room the narrow space in which it operates in the wider universe.

After publishing the only book of philosophy that appeared in his lifetime, Wittgenstein spent the last decades of his life inventing what he called 'language-games', playful scenarios that expose the limitations and falsehoods of language. 'What we are destroying,' he said, 'is nothing but houses of cards and we are clearing up the ground of language on which they stand.'[30]

According to some biographers, Wittgenstein stuttered as a child.[31] I believe he continued to do so at times throughout his life, but there is something unusual in the way in which he did. When you read descriptions of a person who stuttered by those who knew them, a pattern tends to emerge, giving a consistent sense of what they sounded like. But in Wittgenstein's case it is a struggle to find any continuity at all. There are accounts and transcriptions of his speeches and debates in which he seems unable to get the words out. The poet Julian Bell wrote a satirical verse about Wittgenstein saying, 'In every company he shouts us down/And stops our sentence stuttering his own.'[32] Another contemporary recalls his public performances being 'tense and often incoherent'.[33] Yet another: 'He had extreme difficulty in expressing himself and his words were unintelligible to me.'[34] Wittgenstein himself occasionally acknowledges these blocks. In one letter, he complains that his jokes 'get jammed and can't come out'.[35] But on other occasions, Wittgenstein seemed able to speak with great fluency. While stuttering is known for its inconsistencies, this variability is more pronounced in Wittgenstein than most others, raising the possibility that he deployed it as a tactic to draw attention to the limits and failure of language itself. In other words, it was another type of language-game.

'When he started to formulate his view on some specific philosophical problem,' writes one colleague, 'we often felt the internal struggle that occurred in him at that very moment, a struggle by which he tried to penetrate from darkness to light under an intense and painful strain, which was even visible on his most expressive face.'[36] In purely descriptive terms this philosophical struggle is indistinguishable from the appearance of a person wrestling with a block. 'My whole tendency was to run against the boundaries of language,' Wittgenstein wrote. 'This running against the walls of our cage is perfectly, absolutely hopeless.'[37] It is impossible to know for sure to what extent he used his impediment as a way of disrupting philosophical flow, but he did on occasion conflate his attempts at philosophy with the experience of a speech disorder. 'I never more than half succeed in expressing what I want to express,' he once said, 'often my writing is nothing but a form of stuttering.'[38]

I believe that stuttering, whether voluntary or involuntary, became for Wittgenstein a way of doing what he felt many other philosophers tended to neglect. It was a method for interrupting the momentum of linguistic flow and the errors of thought it leads us into, of continually pausing to take stock, reassessing his own ideas and language, then starting again on surer ground.

All speech disorders are disruptive. They break the flow of speech, drawing undue attention to the words we use and the way we say them. More often than not this causes a deep, almost inexplicable, discomfort for both speaker and listener. Normally, this phenomenon is described only in negative terms: it hampers efficient communication, limits public discourse, is a source of embarrassment for all involved. But a neurodiverse account of speech disorders turns this on its head. All those things may be true, but since hyper-fluency is a double-edged sword – able to conceal errors of thinking and prejudice behind pleasant words, as well as enable communication and new ideas – anything that slows it down and subjects it to closer scrutiny has value. In this regard, speech disorders have a social purpose above and beyond the affliction caused in an individual: as a disruptor, a creative spur, a memento of the limits of language or even an agent of resistance against oppression. Until the attitudinal changes that occurred around the time of the First World War, such an idea was implausible, but it has increasingly gained momentum to the point today where Greta Thunberg openly talks about her difficulties with speech as a 'superpower'.[39]

'The stutterer is faithful to human tension every time he talks,' writes American writer David Shields, echoing Wittgenstein, 'only in broken speech is the form of disfluency consonant with the chaos of the world's content. Stutterers are truth-tellers; everyone else is lying. I know it's insane but I believe that.'[40] Joshua St Pierre sees this in more practical terms. 'I genuinely feel that embracing disfluency within our midst changes how we can relate to each other, perhaps in important ways,' he says. 'Disfluency can call a kind of responsiveness from each other that is lacking. It can cultivate new ways of being in the world. At a different level, it's also a pretty strong critique of a lot of these capitalist forces that incite our tongue to speech the whole time.'

I would argue that the societal benefit of speech disorders lies in the same disruptive qualities that those who have them are often most ashamed of. Whether intentionally or not (and mostly the latter), they provide a healthy check on the dangerous errors of thought and action that fluent and hyper-fluent speech can sometimes cause. Along with those productive qualities I have already described – the unacknowledged positive impacts they can have on an individual's personality and creativity – the foundation for a neurodiverse appreciation of speech disorders is evident. Yet it remains far from widely accepted. The challenge lies not so much in describing these phenomena, but in convincing a fluent and prejudiced majority to recognise them. The stakes are high. Bearing in mind the failure of medical science to provide cures (and in some cases even effective diagnoses) of many conditions, neurodiversity presents the best opportunity to reduce the stigma and therefore many of the attendant psychological symptoms for those who have speech disorders. The challenge is to deliver it.

11

Communication Diversity

How do you change the way a society thinks? Shifting mainstream perception is notoriously difficult, particularly when it comes to something as fraught with ideology, custom and prejudice as language. All our instincts lean towards greater regulation rather than relaxation of linguistic performance. There are, after all, hundreds of books on how to speak with greater fluency, always with the threat that without this skill our careers and even our private lives may flounder.

If, as we have seen, there is a general belief that our civilisation is built and depends upon the efficient flow of speech and language, then the unreliability of words is problematic. 'English (or any other language people speak) is hopelessly unsuited to serve as our internal medium of computation,' writes Steven Pinker.[1] He points to its ambiguities, quoting newspaper headlines ('Drunk Gets Nine Months in Violin Case'), and also the fact that the linearity and comparative slowness of speech are almost at cross-purposes with the speed and lateral connectivity of thought.

Speech is particularly prone to sloppiness, partly because it is (mostly) spontaneous and partly because in addition to the ambiguity of individual words there are the additional ambiguities of dialect and intonation. When you put involuntary disfluencies into the mix, like a stutter, aphasia, dysarthria and vocal tics, it appears all the more unreliable. 'There is good reason why so-called laziness in pronunciation is in fact tightly regulated by phonological rules,' Pinker writes, 'and why, as a consequence, no dialect allows its speakers to cut corners at will. Every act of sloppiness on the part of a speaker demands a compensating measure of mental effort on the part of the conversational partner. A society of lazy talkers would be a society of hard-working listeners.'[2] It is assumed that such a society is unwelcome.

We are, it seems, hard-wired to resist variations in speech because we have put so much stock in an instrument that is dangerously unreliable. Yet despite the tendency to police our language, there are occasions when attitudes to linguistic difference or variation have relaxed, and in relatively short time too. By focusing on a couple of these, it is possible to uncover ways in which changing attitudes to speech disorders might be encouraged and achieved.

Martinique is a small island in the Antilles archipelago in the Caribbean. It was settled by the French in the seventeenth century and, aside from some back-and-forth with the British during the Napoleonic Wars, has remained French to this day. Most of the population are descended from the African slaves who were once imported on a mass scale to work the sugar plantations. Those slaves acquired the language of their masters, but in speaking to one another created a hybrid tongue called Creole using French, Carib and African languages, mixed with elements of English, Spanish and Portuguese, as well as inventing entirely new words. A couple of decades ago Martinican Creole was considered by the French, but also many Martinican natives, as it had always been: simply an inferior, mongrel dialect of the mother tongue; a childish babble incapable of carrying abstract thought or communicating knowledge.[3] This way of thinking dominated through to the 1980s, but then, in just a couple of years, Creole suddenly became respected and acknowledged as a linguistic register in its own right.

The seeds of this change date back to 1936 when a young Martinican student called Aimé Césaire returned from several years' study in Paris. He had imbibed the culture of his colonial masters, the excitement of Surrealism and jazz in the cafés and clubs, but also the racism and arrogance he encountered on the streets. He felt dislocated, yearning for the best of both places, but not feeling quite at home in either. He put his thoughts to paper in a prose-poem called *Notebook of a Return to My Native Land,* which captures both the beauty of the island and the subjugation of its people.

Césaire wrote in French, but he also recognised – as Ngũgĩ wa Thiong'o and Dambudzo Marechera would do in coming decades – that the imposition of a master's language is the most effective means of mental colonisation. For Césaire, French was an elegant means of expression but also an implement of domination, even torture. 'Who twists my voice?' he writes in the *Notebook,* describing this predicament. 'Who scratches my voice? Stuffing my throat with a thousand bamboo fangs. A thousand sea-urchin stakes.'[4] In response, Césaire decided, in his own words, to 'bend French': mangling existing words and inventing entirely new ones to say the things about Martinique that dictionary French, a language rooted in Europe, could not capture. It was said by one contemporary that Césaire even had a stutter at this time that began to alleviate with his political and linguistic awakening.[5]

After finishing the *Notebook*, Césaire continued to write but he also sought to liberate Martinique and its people from the worst elements of colonisation. As well as becoming mayor of the capital, he became a schoolteacher, educating the island's children to think for themselves. One of these pupils was a boy called Frantz Fanon. Under Césaire, Fanon learned to distrust the language of France as well as its uniformed authorities, but this also made him more curious about the Creole he and his friends spoke in the streets. Why were Martinicans so free with it in private yet so ashamed of it in public? What psychological impact did the use of two languages, one a subversive play on the other, have on a people?

During the Second World War, Fanon fought with the Free French Forces against Hitler and the Vichy government, but even while risking his life for the liberation of France he encountered appalling racism from those he served. Radicalised by these experiences, he devoted his life to fighting colonialism. He worked for revolutionary movements in both Africa and the Caribbean and provided a philosophical framework for post-colonialism in *Black Skin, White Masks* (1952) and *The Wretched of the Earth* (1961). In the former, Fanon addresses the subjugation of Creole in Martinique. 'The middle class in the Antilles never speak Creole except to their servants.

In school the children of Martinique are now taught to scorn the dialect. One avoids Creolisms. Some families completely forbid the use of Creole, and mothers ridicule their children for speaking it.'[6] Fanon doesn't make a claim for Creole as a creative language in its own right, but he does show that its repression is unsustainable, creating a sense of 'dislocation, a separation' in the hearts of Martinicans. Fanon had taken up the intellectual baton where Césaire left off and he handed it on, in turn, to a younger schoolmate who took this exploration of Creole to a surprising conclusion.

Beginning in the 1970s, Edouard Glissant, a philosopher and poet, argued that Creole was not a mongrel dialect, an inferior subset of French created by people too sloppy to obey the rules. It was a secretive means of communication within plantation life: a *lingua franca* for slaves, using elements of languages familiar to them, that their masters could not understand. When spoken quickly it appeared an 'accelerated nonsense created by scrambled sounds', but was perfectly comprehensible to those using it. As a form of linguistic subterfuge, it was arguably more rather than less sophisticated than the French it toyed with.

Glissant was friends with the French philosophers Deleuze and Guattari and seems to have been influenced by their notion of 'creative stuttering' as a means of resisting the conscription of language to oppressive causes. In *Caribbean Discourse*, published in 1981, Glissant turns on those who try to dismiss Creole. 'You wish to reduce me to a childish babble,' he wrote. 'I will make this babble systematic, we shall see if you can make sense of it.'

Glissant called his study of Creole a 'poetics': the term used for describing linguistic techniques in poetry and literature, but which he defines as 'the implicit or explicit manipulation of self-expression'. By doing so, he elevated its status from street patois to a conscious, organised and authentic language.

In identifying the qualities that synthesise Creole, Glissant articulates a set of techniques for making a dominant, imperial language stutter. He places the objective of 'diversion' as the driving force in its evolution: the need to conceal real meaning from the French administrators, while appearing to dutifully speak a clumsy version of their language. Diversion, in turn, leads to new possibilities: the use of 'ornate expressions and circumlocutions', 'antiphrasis' (using words in a way opposite to their real meaning), 'sudden changes in tone', 'continuous breaks in the narrative', 'asides' and 'the art of repetition'. All of these techniques are familiar to anyone who daily tries to conceal a stutter. According to Glissant, what was considered baby-talk was in fact a form of guerrilla linguistics.

Even as he was writing, a generation of young Martinican novelists used the techniques Glissant prescribed to create a new movement creolising the language of French literature. They called it, simply, Créolité. They put down their thoughts in an essay 'In Praise of Creoleness', which celebrates their language, in opposition to French, as 'an annihilation of false universality, of monolinguism, and of purity'.[7] They both revered Césaire as an intellectual antecedent who discovered the role of the French language in colonialism, as well as the 'bending' of it as a form of resistance, but also criticised him for not going far enough in recognising the beauty and complexity of Creole. In 1992, Patrick Chamoiseau, one of the founders of Créolité, published *Texaco*. It is a powerful and sprawling epic, written in a mixture of French and Creole, that tells the brutal story of the island from the early nineteenth century to his own time. It was quickly recognised as a masterpiece and awarded France's most prestigious literary prize, the Prix Goncourt.

In just a few decades, through organisation, hard work and creative inspiration, a handful of individuals had managed to change the perception of French (or at least the version of it dictated by the Académie française in Paris) from the language of poetry to that of authoritarianism and even oppression, and the language of Creole from degenerate street patois to high art. This, in turn, has contributed to an increasing respect for Creole and recognition of it as a language in its own right. Today, it is studied in books and taught in universities. The concerns about it are no longer a regret that it exists at all, but a fear that in a globalised twenty-first century where Creole is admired rather than ridiculed, it simply becomes assimilated into the dominant languages it once, by necessity, sought to distinguish itself from. Inspired by the lesson of Creole, I wonder if such a similar reversal in perception is possible for speech disorders. Could our reverence for hyper-fluency as the ultimate form of communication diminish, elevating in the process our appreciation of speech disorders, as well as all other forms of verbal and non-verbal communication?

This change has already begun with sign languages. These are forms of expression that use gestures (hand movements, facial expressions and body language) to communicate. They are developed and mostly used by those who are deaf or unable to speak, but they are also learned by those who are part of their lives: family, friends and interpreters. Like spoken language, there are many different forms of Sign across the world and throughout history. Their presence in smaller communities and in historical documents suggest they have always existed.

In the late nineteenth century, the 'oralist' movement sought to eliminate the use of Sign. In the United States, the Scottish inventor Alexander Graham Bell founded the American Association to Promote the Teaching of Speech to the Deaf. Like thousands of others, he believed sign language, something he referred to disparagingly as 'pantomime',[8] should be discouraged, that deaf people shouldn't mix with other deaf people at school or marry them. Whether his views were tempered or exaggerated by the fact that both his mother and his wife were deaf is hard to tell. His greatest invention, the telephone, was a form of oralist communication that many deaf people were (up until recently) unable to use. The success of the oralist movement was profound. By the time of the First World War, around 80 per cent of deaf children were being educated without any access to sign language.[9] Instead, their communication was limited to a poor and reluctant replica of the speech other children performed so naturally.

Gradually, a fightback occurred. In the same way that thinkers like Edouard Glissant revealed the complexity of Creole, a concurrent movement was making a similar case for sign language. In *Sign Language Structure*, published in 1960, the linguist William Stokoe convincingly demonstrated its complexity.[10] Over the ensuing years, he argued that it was not only a sophisticated alternative to spoken language but also a different way of thinking. Many who use Sign describe its physical, three-dimensional and non-linear qualities, in contrast to the linear sentence structures of spoken languages. 'Deafness is a culture and a life,' writes Andrew Solomon, 'a language and an aesthetic, a physicality and an intimacy different from all others.'[11]

In the 1970s, this movement, sometimes called Deaf Pride, gathered momentum and ultimately overturned the supremacy of oralism. One outcome was the formation of Bi-Bi schools (bilingual-bicultural education) where students are taught in Sign with English as a second language. Another, if unexpected, outcome has been the surge in popularity of American Sign Language (ASL) outside of the Deaf community by those exposed to it through personal experience or through seeing signed programmes on television. ASL is now the fifth most taught language in college, practised by as many as two million Americans. As Solomon writes: 'a broad population has been bewitched by the perceived poetry of a physical communication system.'[12]

Deaf culture is not a frictionless utopia. There are those who think Sign should not be assimilated into the mainstream because it is not just a language but the soul and identity of a marginalised community. Meanwhile, the emergence of hearing-enabling cochlear implants, readily available since the 1980s, has been divisive in part because of the extent to which they enable deaf children to 'pass' in mainstream schools and then through professional life at a perceived cost to Deaf culture. But the very fact implants should be so controversial, with many opting out of using it, shows how successful Deaf activism has been.

Mainstream culture has gradually embraced Creole and sign language, yet it is a change that has occurred within my lifetime, even if the origins go back further. It is hardly inconceivable, therefore, that we may in a decade or two think in a similar way about speech disorders. We may recognise the 'ornate expressions and circumlocutions', 'sudden changes in tone', 'continuous breaks in the narrative' and 'the art of repetition' (borrowing the terms Glissant used to describe Creole) within stuttered speech. Likewise, we may celebrate the unpredictability of Tourette's, the economy of dysarthria and alternative communication strategies of aphasia. This is not to whitewash the psychological and physical suffering that can accompany those conditions, in the same way that recognising Creole as a poetic language does not alleviate the pain of colonial subjugation, but it does affirm that there are productive as well as negative ways of thinking about them.

While Creole and Sign are very different forms of communication, there are some similarities in the way the prejudice against them has been reduced, if not entirely overcome. They provide valuable hints for a strategy to shift mainstream perception of speech disorders from stigma to simply different forms of communication. In each case, three things are at play: the behaviour and actions of those within a particular linguistic register;

those, like parents and peers, who encounter them day-in-day-out; and those who do so only occasionally. In the case of Sign, for instance, there was the empowerment and organisation of deaf people; the support of their parents, as well as hearing friends and carers; and then an outside world that, responding to such activism, became more open-minded and ultimately appreciative.

A change in the perception of a minority interest by the mainstream depends, first and foremost, on the activism of that minority. It is the same activism that drove changes in attitude to Creole and Sign. It requires a shift from being a demographic category or percentage (like the million plus in the UK who have a speech disorder) to a collaborative community. Within the UK, the British Deaf Association was formed in 1890, the parents of children with autism formed the National Autistic Society in 1962 (just a couple of decades after it was first diagnosed) and the British Dyslexia Association formed in 1972. These organisations have raised awareness, lobbied the media and government, and enabled the formation of activist groups within their communities. Along with others, they have managed to change the way those conditions are perceived.

Similar associations for those with speech disorders emerged later. The British Stammering Association was founded in 1978, Tourettes Action in 1980 and the Tavistock Trust for Aphasia in 1992 (resulting in the Aphasia Alliance in 2004). Dysarthria is represented separately through organisations like Stroke Association, MND Association, Parkinson's UK and Scope. The closest thing to a single organisation that looks at speech disorders in their entirety is the Royal College of Speech and Language Therapists, although it ultimately represents the professionals who practise speech therapy rather than the individuals who have speech disorders. As Erving Goffman observed back in the 1960s, the 'peculiarity' of speech disorders 'apparently discourages any group formation whatsoever'. Although this has changed, and continues to do so, progress is slow and the handful of organisations in the UK were founded relatively recently and often struggle with funding.

Of course, institutions are just one measure rather than the sum of community activism. There is also what individuals can achieve on their own or with a few allies. Any change in mainstream perception of speech disorders needs to begin with the behaviour of those who have them. The Stuttering Pride movement emphasises the importance of self-empowerment. This means not apologising for stuttering by continually saying 'I'm sorry' or through those symbolic gestures, like the avoidance of eye contact, that imply it. It means eliminating internalised behaviour – word substitution, avoidance of particular situations – in favour of overt stuttering. It means changing the descriptive words used to describe the experience of stuttering: not a speech 'impediment' or even 'disorder' but a 'difference'; not a 'bad' or 'appalling' stutter, but an 'overt' or 'pronounced' one. And it means communicating the benefits and positive insights that come from a speech disorder as well as acknowledging the frustrations it causes. The same principles apply to vocal tics, aphasia and dysarthria. If the many people with speech disorders put as much energy into such overt behaviour as they currently do into avoidance and mitigating tactics, society would be compelled to think about them differently.

Whether as individuals or part of a group or institution, the important thing is to change the public perception of speech disorders. This means challenging wherever possible the popular but pejorative image of the isolated individual with an obscure condition who just can't get their words out. Together, people with speech disorders comprise a vast demographic segment that is full of variation, but also contains more shared characteristics (neurological, psychological and cultural) than has previously been acknowledged. They experience language differently, which means they think differently, and scans show important variations in brain structure. They have a long and overlapping history of misdiagnosis and maltreatment with some shared heroes, like Oliver Sacks, and also villains, like the anally fixated psychoanalyst Sándor Ferenczi. And there is a rich and ongoing tradition of cultural and artistic creation that is either made by or tries to capture the experiences of those with speech disorders: from Lewis Carroll and Samuel Beckett to Edwyn Collins and Jess Thom today. Most of all, they have an acute awareness of the traps and limitations inherent in language that we can all benefit from sharing.

The more people with speech disorders speak out – about their own experiences, about the prejudices surrounding language and the daily reality of discrimination – the more public perception will change. But they cannot be expected to complete this shift on their own. In some cases, they will struggle to change their own perception of themselves let alone that of strangers, for it requires unpicking a lifetime of habit and negative association such as I acquired. A person's behaviour is generally established early in their experience of a speech disorder, determined by how those around them react to it. For those conditions that emerge in childhood, like stuttering or tics, the response of parents and teachers is critical. While for those who develop one later in life, it is the immediate reaction of their family and friends as well as the professionals they are treated by that matters.

Unfortunately, advice for parents of children is conflicted. There are those who recommend acting as quickly as possible if symptoms of disordered speech begin to emerge in their child, intervening in such behaviour before it becomes entrenched. Then there are those who recommend ignoring it for as long as possible unless, or until, it becomes demonstrably upsetting for the child. Finally, there are those – a minority – who consider most intervention nothing more than neurotypical discrimination.[13] This is an ideological issue and each parent must choose their own path, but the priority should be the long-term mental well-being of the child rather than second-guessing the perceived demands of an intolerant society. Increasingly, as schools get better at teaching the benefits of diversity, there are children whose experience of a speech disorder is not quite so bad as it was in my day.

For those conditions which generally emerge later in life, like aphasia or progressive dysarthria, family and peers can help by enabling better communication, starting with themselves. That involves listening better and with greater patience, encouraging non-verbal communication strategies if appropriate, and being mindful of the language they use to describe these changes both with the individual and with their wider social network. Treating it as a sickness raises the possibility of cure but while the symptoms of aphasia can improve, and those of dysarthria may reach a plateau, they are unlikely to be alleviated altogether. It is far better to see it as a great change, fraught with complication and frustration, but one that can also lead to more imaginative and even better communication.

Whether the person in question is a child or adult, the priority should be to provide support and space for them to determine their own response to a disorder rather than unwanted or hasty interference. As parents, partners, friends or carers, we can help by being as matter-of-fact as possible, avoiding both emotive language and labelling before a solid diagnosis is given. It means listening and taking our cue from their signals about whether they want help and what sort of help might be appropriate. Since a bad therapist can worsen a condition, while a good one can alleviate it, and since no cure has ever been found for any speech disorder, it is worth every parent or peer being aware that the personality of the therapist, and the relationship they forge with a client, are probably going to matter more than any treatment or ideology they espouse.

Finally, there are those who neither have a speech disorder nor are intimate with anyone who does. How are they to behave to enable greater acceptance of speech disorders, and why should they bother? I have made the case for the advantages that those with speech disorders bring to our society and culture, but it is also a matter of empathy, whether one believes it is better to be kind than cruel, inclusive rather than divisive. The advice is simple and rooted in common sense:

- Do not avoid communication with somebody who struggles with their speech. Whether it is a person with a mild condition or who communicates through an augmented and alternative communication device, nobody wants to be isolated or avoided because of who they are.
- If they are helped by family or a carer, do not speak to that person as a substitute unless encouraged to do so. It is not up to you whether or when somebody speaks on their behalf.
- Do not be afraid to ask them to repeat themselves. If you find it hard to understand what they are saying, ask again or acknowledge you don't understand. They will be used to it and it is better to establish the terms of engagement than to be ignored.
- Do not speak for them. The temptation is to finish a person's words or sentences. This is literally robbing somebody of their voice and you will probably get it wrong.
- Do not judge or pity. Their experience of language and human society is not inferior to yours but likely to be more nuanced, full of insight that would never have occurred to you because their participation in both is so hard-won.
- Do not discriminate. There are very few experiences or jobs that people with speech disorders are really disadvantaged from

participating in or performing, although it might at times feel easier to dismiss them from the outset.

- Be generous and be brave, lean in rather than lean out. Life is about new experiences and ideas: speech disorders promise both.

The good news is that mainstream perception change is already occurring. Children and young adults with speech disorders seem not to report the same level of bullying and humiliation that older generations describe. The emphasis in speech therapy has shifted from cure at all costs to helping an individual find the level of fluency that is right for them. Workplaces and companies, compelled by anti-discriminatory legislation, are beginning to have policies in place that favour individuals for the work they do rather than the way they speak.[14]

This shift is increasingly reflected and stimulated in the culture around us. Comedy has always been an art form that uses laughter to broach taboo and challenge prejudice. Daniel Kitson and Drew Lynch are both stand-up comics who overtly stutter in their performances, rather than trying to conceal it. There are theatre companies like Jess Thom's Touretteshero and Graeae that showcase work rooted in neurological and physical difference. Lee Ridley (aka Lost Voice Guy), Francesca Martinez and Rosie Jones are all successful comedians who have cerebral palsy and dysarthria of speech. And, at the time of writing, four comedies are on television that feature characters with cerebral palsy, including *Special* and *Speechless* in America and *Don't Forget the Driver* and

Jerk in the UK. There are also those whose work does not revolve around a speech disorder but who have become more open about acknowledging their experience of one. The politician Ed Balls recently opened up about his life-long stutter in his autobiography, and it is questionable whether talented celebrities like actor Emily Blunt or rapper Kendrick Lamar would have been quite so candid about their speech difficulties in a previous age.

I believe that positive mainstream exposure does and will lead to greater acceptance. 'Real exposure to difference is the only way to combat the fear and prejudice that arise out of ignorance and lack of experience,' says Martinez.[15] While Ridley describes how 'for the first time people seem comfortable talking to me, as a disabled person, right from the off'.[16] But there is still a way to go. Broadcast and streaming media are for the most part dominated by hyper-fluent communicators. Our newsreaders, DJs and talk show hosts are expected to have immaculate delivery. After his first day on Radio 4's *Today* programme following his operation for lung cancer, Nick Robinson's voice was ever-so-slightly croaky: something which drew a great deal of attention, even complaints, and for which he felt a need to apologise.

I am convinced that any progress is a trickle-down effect from the success of both neurodiversity and the social model of disability as movements. There is a growing acknowledgement that it is unacceptable to discriminate against difference and that it is the responsibility of society to accommodate rather than vice versa. But it also stems from an increasing appetite for alternative forms of communication to the relentless hyper-fluency that seems to dominate life in the twenty-first century. The current fashion for mindfulness meditation, which is rooted in the non-verbal communication practices of Eastern religions, is partly a craving to silence, if only briefly, the chatter in our minds, mouths and ears. And with many of us increasingly wary of sales patter, political bombast and corporate jargon, language that doesn't flow seems more intriguing than ever before. The so-called Greta Thunberg Effect is simply one example of an ever-increasing appreciation of neurodiversity and alternative forms of communication.

Undoubtedly, these changes are also fuelled by the emergence of social media that has given a voice to disenfranchised communities and individuals like those with autism and dyslexia. It is also inherent in the technology itself. Viewed on one level, platforms like Facebook and Instagram simply mainstream techniques that have been used in augmentative and alternative communication practice for decades. What we once thought was the last resort of the disabled is shown to be a preference for the millions who are happier communing on their devices than in verbal conversation.

In a report from 2015, it was shown that over 72 per cent of eighteen- to twenty-five-year-olds find it easier to put their feelings across in emoji than with words.[17] Something seen by an older generation as a light-hearted punctuation at the end of a text message has emerged as a linguistic system in its own right for the young. Emoji is not an alternative to the word, but a symbolic system that co-exists with it, enabling the easy communication of complex emotions that words alone struggle to capture.

An emoji can reinforce the emotional subtext of a message or deliver irony, humour or poignancy by contradicting it, while a combination of emojis can allow everyone to communicate difficult emotions that were once limited to a handful of poets adept at the sonnet form. 'Emoji is the fastest growing form of language in history based on its incredible adoption rate and speed of evolution,' claims Professor Vyv Evans from Bangor University.[18] Only recently, the Oxford Dictionaries announced their word of the year was U+1F602 (or 'face with tears of joy'). Like any emerging technology, we can only guess what this new language will ultimately mean for communication, but it certainly reduces the importance of the spoken word.

What are we to call this new movement that affects our understanding of speech and communication, that draws on not only the rise of neurodiversity, but also the advance of technology and a broader acceptance of alternative forms of communication? I spent months searching for the right term, then stumbled upon it while listening to a recent podcast encounter between two highly creative individuals with speech disorders. The first is a poet and rapper from Essex called Scroobius Pip. His career began in the early years of the new millennium, when a recording of a poem about his stutter drew the attention of radio producers and the music industry.

In 2013, the radio station XFM signed Scroobius Pip to host a weekly hip hop and spoken word programme. Some thought this was brave as Pip's stutter, which had made his name, hadn't gone away. Because the show was pre-recorded there was the option to edit his speech and Pip found himself with a golden opportunity that many with speech disorders strive for: he could present himself to the world as fluent. Yet Pip's stutter was part of who he was and it had inspired the work that first brought him attention. So, as he tells me when we meet in an east London café, he made a momentous decision. 'Look, if I really get stuck on something,' he said to the producer, 'I might start again. But other than that, I don't want you editing out my stutter. I don't want us doing retakes and retakes and retakes. As long as we can get everything into the hour show, as long as we're not missing any songs because of my stutter, it's all good with me.' Pip's show was the only programme I'm aware of anywhere in the mainstream media presented by a person who conspicuously stutters.

Then Pip went independent. He started his own label, which he called Speech Development Records, because he wanted something he could rhyme with 'speech impediment'. And he launched a weekly podcast called *Distraction Pieces*, in which he interviews artists and celebrities like Russell Brand and Killer Mike. It has built up a steady following: thirty million downloads in all. What is immediately noticeable about the podcast is that the host stutters a lot. 'A small radio slot I can probably get through without stuttering,' he tells me, 'but a ninety-minute podcast – there's going to be stuttering and I'm not going to edit it out.' He doesn't see himself as an activist though. He's not proud of his stutter, but simply recognises it as part of who he is. 'I think of it like an accent,' he says. 'I do have people hit me up and say it's so bold or so inspirational, but the fact is I've never considered it. I've just thought: I want to do a show. I haven't thought how empowering this is.'

Over time, Pip has come to think more about the values he stands by. 'I was coming up to my 200th episode,' he says. 'I had the option of Russell Brand coming back on. I had the option of a few huge names. And I realised that a huge name will get a lot of listeners at any time. The importance of the 200th episode is showing what the *Distraction Pieces* podcast is.' So he invited Jess Thom onto the show. The podcast is long (over eighty minutes) not because their speech slows them down but because they have so much to say. 'I realised that we've got the opposites,' Pip tells me. 'She can't stop things coming out and I can't make things come out. In that moment we realised that there's not been many conversations listened to by tens of thousands between people who don't speak in the traditional manner, who have some kind of restriction over what they say.'

In the podcast, there's a moment talking on this exact theme when Thom suddenly shouts out, 'communication diversity, motherfuckers!' The swear word makes it sound like a tic, yet the phrase 'communication diversity' perfectly sums up what is going on in episode 200 of *Distraction Pieces*: a programme, impossible only ten years ago, that gives space to two individuals with speech differences to communicate with a large audience, not only without concealing but actively celebrating those differences. It's not a phrase in common usage so when I ask Thom where it came from, she says that she doesn't know and it doesn't really matter. Whether or not the phrase 'communication diversity, motherfuckers!' was a tic or not doesn't detract from its potency but adds to it. The term itself might be the product of the sort of speech differences it seeks to accommodate.

Communication Diversity

Communication diversity recognises that, despite the tendency for fluency-prejudice throughout society, there are many modes of speaking and communicating. We should champion them all: not only because it makes life more rewarding for the many millions who do communicate in different ways, but because it is how we hold in check the normative tyranny of fluency, the errors of thinking it can lead to and the unquestioning trust we have in its operation. Communication diversity isn't against language and fluency, and it's not limited to speech disorders, it simply seeks to put all registers of speech and linguistic usage into what Kenyan writer Ngũgĩ wa Thiong'o described to me as a 'network of equal give and take and not as hierarchies of power'.

Embracing this way of thinking has been difficult because of an immense cultural bias stacked against it. It means accepting that fast, fluent, eloquent speech is only one form of good communication. It means embracing ambiguity, being willing to slow down, to listen properly rather than depending on the use of a verbal shorthand that both eases communication and also allows one not to concentrate too closely. Most of all, it means accepting that people whose speech is distorted, fragmented, slow or even entirely absent are not necessarily inferior communicators; they just depend on a degree of engagement and attention we are not used to giving. Steven Pinker's warning about the 'society of hard-working listeners' that must result from 'laziness in pronunciation' is another person's utopia when compared to the errors and confusions that arise daily, not from lazy talking but lazy listening.

The rise of communication diversity is partly a reflection of a confident society that is more willing to encompass and celebrate difference, but it also arises from an increased acknowledgement of the limitations of speech. Yes, it is one of humanity's great creations, a tool that has enabled us to achieve extraordinary things. But where the Victorians unquestioningly praised the benefits of rhetoric and hyper-fluency against the supposed babble of different linguistic registers, there is enough precedence now in acknowledging their limitations. This change, a form of positive disillusionment, is impossible to separate from the failure of that eighteenth-century (or Enlightenment) project of Progress, in which language and technology were seen as the key to a better future. For while that same progress has brought great freedoms and well-being, it has also brought us social and environmental disequilibrium on a scale unprecedented in human history. While we continue to recognise that words help us, we also recognise the extent to which they have failed us too.

I began by asking why King George VI's ability to speak fluently seemed to matter so much. I have tried to show that it has very little to do with the inherent shamefulness of a tendency to get stuck on certain words, but is really about the cracks that run through our civilisation. By putting too much at stake – our knowledge, our institutions and values – in the exercise of fluent language, we put ourselves, as a society, in the ludicrous position where the King's stutter seemed to undermine everything we stood for. The notion is, of course, absurd, yet its very absurdity shows how ill-founded and insupportable our trust in that single mode of hyper-fluent linguistic usage really is. Seen objectively, the King's stutter did not matter, and the conviction that it demeaned not only him, who tried so hard to conceal it, but also – more substantially – the whole society of 1940s Britain, with its values, beliefs and customs, that made him feel he ought to. We need to ensure that this can't and doesn't happen again.

The key to achieving this lies simply in telling stories. We know the familiar ones: the neurotic stutterer, the mentally impaired aphasiac, the obscene Tourettic and the sorry faces of dysarthria. Now it's time to tell the real ones, using whatever means we have available. I spent a lifetime doing my best to hide my stutter. On the occasions when it was unavoidably present, I made it a laughing point, even exaggerated it, as if my only value as a person who stutters was to provide a little amusement for others, apologising as I did so. I played the clown, but I will not do so any more.

Epilogue: Out of the Mouth Trap

One evening in 2009, I found myself in a state of extreme anxiety standing in the dark and drizzle outside Holborn station. The entrance was fanned by a crowd of commuters trying to squeeze their way out of the rain, through the ticket barriers and down to the platforms below. No such escape was available for me though. I had a task to do; one that filled me with more fear than almost anything I had ever done.

For some time, I procrastinated, finding endless excuses, mostly to do with the fanciful notion that those passing me were the wrong sort of people. But finally, fed up with being wet and cold, and disgusted by my own inaction, I made my move. Lurching towards the nearest passer-by, I said, 'Excuse me!' The man stopped. He was a tired, harassed-looking commuter who glanced rather regretfully towards the station entrance but turned nonetheless. I looked him in the eye and smiled. 'Can you tell me the way to ...' And then I blocked on a hard C sound. I felt my top lip quiver, my nostrils flare, my chin strain. I hit that 'Co-' sound five times, never breaking eye contact, and then released: 'Covent Garden?'

To my surprise, the weary businessman didn't laugh

or sneer, but simply gave me directions and hurried on without another word. Apparently, he just wanted to get home. Emboldened, I repeated the exercise on another passer-by, then another. I did it seven times in total. I'd like to say they all exhibited the same disinterested reaction as the first, but one gave a quizzical smirk as I blocked and another laughed outright. I then returned to the strip-lit classroom barely a hundred metres away, where ten people with stutters like mine were gradually regrouping. This was the City Lit course for Interiorised Stammering. We were all people who had constructed our lives around avoiding the act of stuttering for the simple reason that, on a good day, we could just about manage to do so. As long as we didn't say certain words, of course, and avoided particular situations, even particular people. We had just made our first out-of-the-classroom foray in the technique of voluntary stuttering.

I am ashamed to say that despite a lifetime of stuttering and avoiding stuttering, I had never looked anyone in the eyes at the crucial moment. Like the Three Sillies of the old English folk story, who live in fear of an axe stuck in the ceiling falling on them without it ever occurring to them that they might reach up and remove it, I had spent my life fearing the disgusted reaction people must have to my stutter without ever looking to see whether it was true. This was behaviour ingrained since childhood and, as a consequence, I had also never really thought about my stutter other than as a revolting trait I needed to conceal. I was thirty-four years old and knew next to nothing about the very phenomenon that had consumed more mental energy than almost anything else in my life. I had no idea what might cause stuttering, how many others experienced it, the different ways it had been treated or how it related to other speech disorders.

That night was a turning point. When I looked that commuter in the eyes, I also looked for the first time at my stutter itself as something I might try to understand rather than hide from. Almost immediately, I began reading into it, beginning with what ultimately forms the first two chapters of this book: an anatomy of the main speech disorders. The ensuing chapters more or less follow, with a little retrospective tidying up, the process of discovery and self-understanding I embarked upon. But there is another story not captured in these pages. In the ten years since that night, stuttering has ceased to be a significant presence in both my spoken and unspoken use of language.

The reason for this is, I think, a perfect storm of elements. There is, of course, the impact of the City Lit course and the excellent speech therapists I was lucky enough to encounter. While endlessly emphasising that there is no cure for stuttering, that they could only help me change my attitude to it, they also made me fear it less, with the outcome that I did it less. While I have met some who are against speech therapy, it changed my life. And I see that reflected in the accounts of others. 'This profession is certainly not given enough credit,' writes Grace Maxwell when describing the impact that speech therapy had on her husband's aphasia. 'The best therapists turn around lives that appear to be wrecked.'[1]

There are other, more prosaic, changes that probably played a role. I was getting older for a start. We still don't know why conditions like stuttering often alleviate

302

with age. It might be about acceptance, it might be about changes in the brain. After all, if the problem is to do with neurological wiring unable to keep up with the activity of different parts of the brain, maybe the brain overall simply slows down a little and that connecting matter finds it easier to keep pace. Then there was my job. I accepted an irresistible promotion that meant I couldn't avoid public speaking any longer. Within a couple of years, I was readily and regularly participating in talks, interviews and debates often on air or in front of large groups of people. I feared it dreadfully at first, neurotically preparing and over-preparing for days, but you can only maintain that level of fear, with the requisite preparation time to compensate, for so long. I learned to speak impromptu. I found that along with the fear was a seed of pleasure which flowered and grew with each event until the anxiety and enjoyment complemented one another. While I still block on words much more when public speaking (and, I should add in self-admonishment, still resort to the sort of word substitution I am trying to avoid), it generally doesn't prevent me from saying what I have to say.

My fluency also increased through researching this book. I became fascinated by those, like Lewis Carroll, Edwyn Collins and Jess Thom, who found a creative outlet for their speech conditions. I felt proud to be associated with what I increasingly saw as a historical and cultural tribe: people with a fundamentally different experience of language going back as far as records allow. Most of all, I found myself deeply moved and inspired by the bravery of those I met: people for whom every sentence was imbued with disordered speech, who in the process of coming out had wrestled with appalling self-loathing, real rather than perceived social discrimination and who in some cases even considered suicide. My own experience of stuttering, mostly interiorised, felt petty in comparison. I am particularly inspired by the younger generation who, still in their teens or in their twenties, openly identify as people with speech disorders and manage to recognise and celebrate the distinctiveness of their speech while also acknowledging the pain it has caused them.

There's a final reason why I think I am more fluent, yet also identify more strongly as a person who stutters than ever before. Shortly after completing the City Lit course, I met a woman whose experience of stuttering was similar to mine. Like many in his time, James Hunt believed people who stutter should avoid one another because it is a form of imitative behaviour and, as a rule of thumb, they still do. But I think it's one thing he got truly wrong. The only way people with speech disorders can change the wider perception of their conditions is by working together. Not only did Constance stutter but most of her family did, simply as my mother had done. Finding myself part of an extended family of people who stutter, although with great variation, was a joy.

In the early years of our marriage, Constance and I found ourselves having to do a lot of public speaking for work. It was a new experience for both of us. We would spend hours rehearsing presentations at one another, which would generally begin with a giant block followed by moments of verbal collapse throughout. We were trying to pass for fluent and by the time the talk came most difficult words had been carefully ironed out. I think the whole process made us less self-conscious and we prepare less intensely now. If we stutter, that's fine. And, of course, just that mental acceptance means we do it less.

At the time of writing, our two children are learning to speak. With so many people who stutter in the family, there is a high chance one of them will too. We find ourselves listening carefully, although we hope not obsessively or neurotically, to their speech development. As a proud father, I keep a track of our eldest's words: mama, papa, caca, goh-goh (meaning both 'crocodile' and 'helicopter'), pah-pah (pasta). All repetitive words like the 'bah-bah' of 'barbarians'; all indistinguishable from what might, in other cases, be a stutter. If learning to speak is a form of stuttering, when does it become considered a 'problem'? And if one of our children does stutter, will we take them to speech therapy, and at what point?

As one of the many born with a speech disorder, I dream, for future generations more than anything else, of their acceptance. But there is an unexpected risk here. Once the shame and the coping mechanisms disappear, once the fear of language and the need to use it differently are rendered obsolete, will speech disorders still have the same creative and productive characteristics I have identified or will they diminish? I want to preserve rather than eliminate such qualities. The solution, I think, is to go one step further. As well as accepting speech disorders, let us all distrust language a little more: vigilant for the delusions and misuses it is put to; more open to experimentation, trying unusual words, strange sentence structures, and resisting rather than striving for more consistent use. Rather than seeking for better inclusion of people with speech disorders in society, let's instead seek to make society use language more like they do. We will, I think, be more tolerant, creative and wiser for it.

Notes

Introduction: The King and I

1. For my account of George VI's youth and coronation I have drawn predominantly from *George VI: The Dutiful King* by Sarah Bradford (Penguin, 2011).
2. Alan and Irene Taylor, *The Assassin's Cloak: An Anthology of the World's Greatest Diarists* (Canongate Books, 2008).
3. Simon Garfield, *Our Hidden Lives: The Remarkable Diaries of Postwar Britain* (Ebury Press, 2005).
4. Cecil Beaton, *The Happy Years: Diaries 1944–48* (Weidenfeld & Nicolson, 1972).
5. David Kynaston, *Austerity Britain, 1945–1951 (Tales of a New Jerusalem),* (Bloomsbury Publishing, 2008).
6. Johannes von Tiling, 'Listener Perceptions of Stuttering, Prolonged Speech, and Verbal Avoidance Behaviors', *Journal of Communication Disorders* (March/April 2011).
7. www.rcslt.org
8. Mark L. Knapp, Judith A. Hall and Terrence G. Horgan, *Nonverbal Communication in Human Interaction,* 8th Revised Edition (Wadsworth Publishing, 2013).

1. Maladies of Speech

1. Gitogo makes a cameo appearance in *A Grain of Wheat* (1967), referred to by name as a deaf mute who was killed by the British. In *Petals of Blood* (1977), all the major characters

307

struggle with inarticulacy, often speaking in broken sentences or breaking off halfway through. Nyakinyua's husband is unable to recount a vision because 'something always blocked him, his throat, in the beginning of telling it, and could not continue.' In *Devil on the Cross* (1980), many of the characters 'stutter like babies' when speaking Gikuyu but are fluent in foreign languages.

2. John A. Tetnowski and Kathy Scaler Scott, 'Fluency and Fluency Disorders', in Jack C. Damico, Nicole Müller and Martin J. Ball (eds), *The Handbook of Language and Speech Disorders* (Wiley-Blackwell, 2013).

3. Marcel Wingate, 'Recovery from Stuttering', *Journal of Speech and Hearing Disorders* (August 1964).

4. David Crystal, *How Language Works: How Babies Babble, Words Change Meaning and Languages Live or Die* (Penguin, 2007).

5. Sarah Bradford, *George VI: The Dutiful King* (Penguin, 2011).

6. Mary Robertson and Andrea Cavanna, *Tourette Syndrome*, 2nd Edition (Oxford University Press, 2008)

7. Ibid.

8. Oliver Sacks, *An Anthropologist on Mars* (Alfred A. Knopf, 1995).

9. Gill Edelman and Robert Greenwood, *Jumbly Words, and Rights Where Wrongs Should Be: The Experience of Aphasia from the Inside* (Far Communications, 1992).

10. Henry Head, *Aphasia and Kindred Disorders of Speech*, Volume 2 (Cambridge University Press, 1926).

11. Chris Code, 'Aphasia', in Jack C. Damico, Nicole Müller and Martin J. Ball (eds), *The Handbook of Language and Speech Disorders* (Wiley-Blackwell, 2013).

12. Gill Edelman and Robert Greenwood, *Jumbly Words, and Rights Where Wrongs Should Be: The Experience of Aphasia from the Inside* (Far Communications, 1992).

13. Ibid.

14. Mona Greenfield and Ellayne S. Ganzfried, *The Word Escapes Me: Voices of Aphasia* (Balboa Press, 2016).
15. www.parkinsonsnewstoday.com/parkinsons-disease-statistics
16. www.cerebralpalsy.org.uk
17. www.mndassociation.org
18. Patrick J. Bradley, 'Voice Disorders: Classification', in *Otorhinolaryngology–Head and Neck Surgery* (Springer, 2010).
19. www.selectivemutism.org.uk
20. Ashley John-Baptiste, 'Selective mutism: "I have a phobia of talking"', *BBC News* (15 July 2015).
21. Greta Thunberg, TEDxStockholm talk (24 November 2018).
22. Office for National Statistics
23. T. Benke, C. Hohenstein, W. Poewe, B. Butterworth, 'Repetitive Speech Phenomena in Parkinson's Disease', *Journal of Neurology, Neurosurgery and Psychiatry* (September 2000).

2. The Mouth Trap

1. All quoted in Leon Edel, *Henry James: A Life* (HarperCollins, 1985).
2. Ibid.
3. Ibid.
4. Ibid.
5. Ibid.
6. Letter from Virginia Woolf to Violet Dickinson, 25 August 1907, in Nigel Nicolson and Joanne Trautmann (eds), *The Letters of Virginia Woolf, Volume One: 1888–1912* (Houghton Mifflin, 1977).
7. Leon Edel, *Henry James: A Life* (HarperCollins, 1985).
8. Edith Wharton, *A Backward Glance* (Appleton-Century Company, 1934).
9. Joseph Sheehan, *Stuttering: Research and Therapy* (Harper and Row, 1970).
10. David Crystal, *How Language Works* (Penguin, 2007).
11. Francesca Martinez, *What the **** is Normal?!* (Virgin Books, 2015).

12. Eugene Frederick Hahn (ed.), *Stuttering: Significant Theories and Therapies* (Stanford University Press, 1956).

13. Wendell Johnson, 'The Indians Have No Word for it: I. Stuttering in Children', *Quarterly Journal of Speech* (1944).

14. John A. Tetnowski and Kathy Scaler Scott, 'Fluency and Fluency Disorders', in Jack C. Damico, Nicole Müller and Martin J. Ball, *The Handbook of Language and Speech Disorders* (Wiley-Blackwell, 2013).

15. Mary Tudor, 'An experimental study of the effect of evaluative labelling on speech and fluency' (MA thesis, University of Iowa, 1939).

16. Franklin H. Silverman, 'The "Monster" Study', *Journal of Fluency Disorders* (June 1988).

17. Darcey Steinke, 'My Stutter Made Me a Better Writer, *New York Times* (6 June 2019).

18. Annual Report of Tourette Association of America 2018.

19. Gill Edelman and Robert Greenwood, *Jumbly Words, and Rights Where Wrongs Should Be: The Experience of Aphasia from the Inside* (Far Communications, 1992).

20. Elizabeth Jordan, 'Henry James at Dinner', *Mark Twain Quarterly* (Spring 1943).

21. Ben Brown, 'Ben Brown's Story', Tourette Association of America website.

22. Jon Palfreman, *Brain Storms: The Race to Unlock the Mysteries of Parkinson's Disease* (Rider Books, 2015).

23. Michael J. Fox, *Lucky Man: A Memoir* (Hyperion, 2002).

24. James Hunt, *Stammering and Stuttering, Their Nature and Treatment* (Longman, Green, Longman and Roberts, 1861).

25. Gill Edelman and Robert Greenwood, *Jumbly Words, and Rights Where Wrongs Should Be: The Experience of Aphasia from the Inside* (Far Communications, 1992).

26. Chris Code, 'Aphasia', in Jack C. Damico, Nicole Müller and Martin J. Ball (eds), *The Handbook of Language and Speech Disorders* (Wiley-Blackwell, 2013).

27. Jane Fraser (ed.), *Do You Stutter: A Guide for Teens* (Stuttering Foundation of America, 1987).

3. Talking Culture

1. Daniel Everett, *Language: The Cultural Tool* (Pantheon Books, 2012).
2. Brian Goldstein and Ramonda Horton-Ikard, 'Diversity Considerations in Speech and Language', in Jack C. Damico, Nicole Müller and Martin J. Ball (eds), *The Handbook of Language and Speech Disorders* (Wiley-Blackwell, 2013).
3. John Colapinto, 'The Interpreter', *The New Yorker* (16 April 2007).
4. Brian Goldstein and Ramonda Horton-Ikard, 'Diversity Considerations in Speech and Language', in Jack C. Damico, Nicole Müller and Martin J. Ball (eds), *The Handbook of Language and Speech Disorders* (Wiley-Blackwell, 2013).
5. Victoria Glendinning, *Elizabeth Bowen: A Biography* (Anchor, 2006).
6. Emily C. Bloom, *The Wireless Past: Anglo-Irish Writers and the BBC, 1931–1968* (Oxford University Press, 2016).
7. Howard I. Kushner, *A Cursing Brain? The Histories of Tourette Syndrome* (Harvard University Press, 1999).
8. Ibid.
9. Erving Goffman, *Stigma: Notes on the Management of Spoiled Identity* (Prentice-Hall, 1963).
10. Suetonius, *De Vita Caesarum* (*Lives of the Caesars*), *circa* AD 121.
11. Seneca the Younger (attributed), *Apocolocyntosis*, *circa* first century AD.
12. Tacitus, *Annales*, Book 3, Chapter 12, *circa* first century AD.
13. Erving Goffman, *The Presentation of Self in Everyday Life* (Doubleday, 1956).
14. The popularising of the term Received Pronunciation is credited to the second edition of Daniel Jones's *English Pronunciation Dictionary* (1924). In the first edition of 1917, he used the

term 'public school English'. 'Standard Southern British' is cited in *Handbook of the International Phonetic Association* (Cambridge University Press, 1999).

15. John Hendrickson, 'What Joe Biden Can't Bring Himself To Say', *The Atlantic* (Jan/Feb 2020).

16. Ibid.

17. Lee Ridley, aka Lost Voice Guy, *I'm Only in it for the Parking: Life and Laughter from the Priority Seats* (Bantam Press, 2019).

18. Alessandro Duranti, *Linguistic Anthropology* (Cambridge University Press, 1997).

19. David Crystal, *How Language Works: How Babies Babble, Words Change Meaning and Languages Live or Die* (Penguin, 2007).

20. William Jordan, 'Afraid of Heights? You're Not Alone' (YouGov, 20 March 2014).

21. Quoted in Ronald Carter, *Language and Creativity: The Art of Common Talk* (Routledge, 2004).

22. Steven Connor, *Beyond Words: Sobs, Hums, Stutters and Other Vocalizations* (Reaktion Books, 2014).

23. John A. Tetnowski and Kathy Scaler Scott, 'Fluency and Fluency Disorders', in Jack C. Damico, Nicole Müller and Martin J. Ball (eds), *The Handbook of Language and Speech Disorders* (Wiley-Blackwell, 2013).

24. Mark L. Knapp, Judith A. Hall and Terrence G. Horgan, *Nonverbal Communication in Human Interaction*, 8th Revised Edition (Wadsworth Publishing, 2013).

25. Albert Mehrabian, *Nonverbal Communication* (Transaction Publishers, 1972).

4. The Tyranny of Fluency

1. Chris Anderson, *TED Talks: The Official TED Guide to Public Speaking* (Hodder and Stoughton, 2018).

2. Susan Cain, *Quiet: The Power of Introverts in a World That Can't Stop Talking* (Penguin, 2013).

3. Victor Klemperer, *The Language of the Third Reich* (Athlone Press, 2000).

4. Ronald B. Adler, George Rodman and Athena du Pré, *Understanding Human Communication*, 13th edition (Oxford University Press, 2016).

5. Andrew Solomon, *Far from the Tree* (Vintage, 2014).

6. Ronald B. Adler, George Rodman and Athena du Pré, *Understanding Human Communication*, 13th edition (Oxford University Press, 2016).

7. Patrick Leigh Fermor, *A Time to Keep Silence* (Queen Anne Press, 1953).

8. Christian Laes, 'Silent History? Speech Impairment in Roman Antiquity', in Christian Laes, C.F. Goodey and M. Lynne Rose (eds), *Disabilities in Roman Antiquity* (BRILL, 2013).

9. Daniel Defoe, *The Fortunes and Misfortunes of the Famous Moll Flanders* (London, 1722).

10. Eugenia Stanhope (ed.), *Letters Written by the Late Right Honourable Philip Dormer Stanhope, Earl of Chesterfield, to his Son, Philip Stanhope, Esq.* (London, 1800).

11. Elizabeth Foyster, '"Fear of Giving Offence Makes Me Give the More Offence": Politeness, Speech and Its Impediments in British Society, c.1660–1800', *Cultural and Social History* (September 2018).

12. Quoted in Susan Cain, *Quiet: The Power of Introverts in a World That Can't Stop Talking* (Penguin, 2013).

5. A Muted History

1. Chris Code, 'Aphasia', in Jack C. Damico, Nicole Müller and Martin J. Ball (eds), *The Handbook of Language and Speech Disorders* (Wiley-Blackwell, 2013).

2. Paul Broca's original statements are recorded in editions of *Bulletin de la Société Anatomique de Paris* (1861). I have used translations from Paul Eling (ed.), *Reader in the History of Aphasia: From Franz Gall to Norman Geschwind* (John Benjamins Publishing Company, 1994).

3. James Hunt, *Stammering and Stuttering, Their Nature and Treatment* (Longman, Green, Longman and Roberts, 1861).

4. J.F. Dieffenbach, *Memoir on the Radical Cure of Stuttering* (Samuel Highley, 1841).

5. Elizabeth Foyster, '"Fear of Giving Offence Makes Me Give the More Offence": Politeness, Speech and Its Impediments in British Society, c.1660–1800', *Cultural and Social History* (September 2018).

6. Erasmus Darwin, *Zoonomia, or the Laws of Organic Life* (London, 1794).

7. Allan Ropper and B.D. Burrell, *How the Brain Lost Its Mind: Sex, Hysteria and the Riddle of Mental Illness* (Atlantic Books, 2020).

8. Howard I. Kushner, *A Cursing Brain? The Histories of Tourette Syndrome* (Harvard University Press, 1999).

9. Sigmund Freud, *The Psychopathology of Everyday Life* (Berlin, 1904).

10. This is not the same as the 'psychical' approach of James Hunt for, while Hunt emphasised the importance of negative associations and mental habits in stuttering, these were conscious and easily identifiable rather than repressed.

11. Sigmund Freud, Sándor Ferenczi, Karl Abraham, Ernst Simmel and Ernest Jones, *Psycho-Analysis and the War Neuroses* (The International Psycho-Analytical Press, 1921).

12. I. Peter Glauber, 'Freud's Contributions on Stuttering: Their Relation to Some Current Insights', *Journal of the American Psychoanalytic Association* (April 1958).

13. Sándor Ferenczi, *Thalassa: A Theory of Genitality* (The International Psycho-Analytical Press, 1924).

14. Otto Fenichel, *The Psychoanalytic Theory of Neurosis* (W.W. Norton and Company, 1945).

15. Isador Coriat, *Stammering: A Psychoanalytic Interpretation* (Nervous and Mental Disease Publishing Company, 1927).

16. Sándor Ferenczi, 'Psycho-analytical Observations on Tic', *International Journal of Psycho-Analysis* (March 1921).

17. Margaret S. Mahler, 'A Psychoanalytic Evaluation of Tic in Psychopathology of Children: Symptomatic Tic and Tic Syndrome' (1949), in *The Selected Papers of Margaret Mahler*, Volume 1 (Aronson, 1979).

18. Margaret S. Mahler, 'Outcome of the Tic Syndrome' (1946), in *The Selected Papers of Margaret Mahler*, Volume 1 (Aronson, 1979).

19. L.S. Jacyna, *Lost Words: Narratives of Language and the Brain, 1825–1926* (Princeton University Press, 2000).

20. Nicholas Mosley, *Beyond the Pale: Sir Oswald Mosley and Family* (Secker & Warburg, 1983).

21. Louise Robison Kent, 'Carbon Dioxide Therapy as a Medical Treatment for Stuttering', *Journal of Speech and Hearing Disorders* (August 1961).

22. Howard I. Kushner.

23. Norman Geschwind, 'Disconnexion Syndromes in Animals and Man', *Brain* (June 1965).

24. Arthur K. Shapiro and Elaine Shapiro, 'Treatment of Gilles de la Tourette's Syndrome with Haloperidol', *British Journal of Psychiatry* (March 1968).

25. Howard I. Kushner.

26. Transcript of interview with Oleh Hornykiewicz conducted by Barbara W. Sommer in Toronto, Canada, 9 February 2007.

6. Unfinished Stories

1. The stuttering Ken in *A Fish Called Wanda* was played by Michael Palin who a few years later gave his name, and a great deal of his time, to the Michael Palin Centre for Stammering Children. In an article for the *Telegraph* in 2011, Palin argued that although the role of Ken was considered cruel by some people who stuttered, others were delighted with it because it got people talking more openly about stuttering.

2. Anne Woodham, 'How My Son Lost the Edge of My Wretched Tongue', the *Guardian* (15 February 1991).

3. Anne Woodham, 'In a Manner of Speaking', *Good Housekeeping* (December, 1987).
4. Ibid.
5. Anna Craig-McQuaide, Harith Akram, Ludvic Zrinzo and Elina Tripoliti, 'A Review of Brain Circuitries Involved in Stuttering', *Frontiers in Human Neuroscience* (2014).
6. Deanne J. Greene, Bradley L. Schlaggar and Kevin J. Black, 'Neuroimaging in Tourette Syndrome: Research Highlights from 2014–2015', *Current Developmental Disorders Reports* (December 2015).
7. Gregory J. Synder, 'The Existence of Stuttering in Sign Language and Other Forms of Expressive Communication: Sufficient Cause for the Emergence of a New Stuttering Paradigm?', *Journal of Stuttering, Advocacy and Research* (January 2009).
8. T.J. Murray, P. Kelly, L. Campbell, K. Stefanik, 'Haloperidol in the treatment of stuttering', *British Journal of Psychiatry* (April 1977).
9. M. Boldrini, M. Rossi, G. F. Placidi, 'Paroxetine Efficacy in Stuttering Treatment', *International Journal of Neuropsychopharmacology* (September 2003).
10. Howard I. Kushner, *A Cursing Brain? The Histories of Tourette Syndrome* (Harvard University Press, 1999).
11. L.S. Jacyna, *Lost Words: Narratives of Language and the Brain, 1825–1926* (Princeton University Press, 2000).
12. Marian C. Brady, Helen Kelly, Jon Godwin, Pam Enderby and Pauline Campbell, 'Speech and Language Therapy for Aphasia Following Stroke', *Cochrane Library* (1 June 2016).
13. Sarah Johnson, 'Nick Robinson: I Never Thought I'd Get My Speech Back', the *Guardian* (28 December 2016).
14. Grace Maxwell, *Falling and Laughing: The Restoration of Edwyn Collins* (Ebury Press, 2010).
15. Nicholas Mosley, *Beyond the Pale: Sir Oswald Mosley and Family* (Secker & Warburg, 1983).

16. Lee Ridley, aka Lost Voice Guy, *I'm Only in it for the Parking: Life and Laughter from the Priority Seats* (Bantam Press, 2019).

7. Extraordinary Minds

1. A person with agnosia struggles to interpret sensations and thereby recognise things; someone with hemispatial neglect loses awareness of one side of vision; someone with somatoparaphrenia may deny the existence of a limb or part of the body.

2. Oliver Sacks, 'The Twins' and 'The Autist Artist', Chapters 23 and 24 of *The Man Who Mistook His Wife for a Hat* (Gerald Duckworth, 1985).

3. Oliver Sacks, 'Witty Ticcy Ray', Chapter 10 of *The Man Who Mistook His Wife for a Hat*.

4. Oliver Sacks, 'The President's Speech', Chapter 9 of *The Man Who Mistook His Wife for a Hat*.

5. Oliver Sacks, *An Anthropologist on Mars* (Alfred A. Knopf, 1995).

6. Landmark reports include Hans Asperger, 'Die Autistischen Psychopathen im Kindesalter' (1944) and Leo Kanner, 'Autistic Disturbances of Affective Contact' (1943).

7. The notion of autism being caused by a 'lack of maternal warmth', effectively leaving children 'in refrigerators which did not defrost', originated in Leo Kanner, 'Problems of Nosology and Psychodynamics in Early Childhood Autism', *American Journal of Orthopsychiatry* (1949).

8. Jason J. Wolff, Suma Jacob and Jed T. Elison, 'The Journey to Autism: Insights from Neuroimaging Studies of Infants and Toddlers', *Development and Psychopathology* (May 2018).

9. Temple Grandin, *Thinking in Pictures: And Other Reports from My Life with Autism* (Random House, 1996).

10. Ron Suskind, *Life, Animated: A Story of Sidekicks, Heroes, and Autism* (Kingswell, 2014).

11. Steve Silberman, *NeuroTribes: The Legacy of Autism and the Future of Neurodiversity* (Avery Publishing, 2015).

12. Maryanne Wolf, 'Dyslexia and the Brain That Thinks Outside the Box', *Dyslexia Review* (2008).

13. Judy Singer, *NeuroDiversity: The Birth of an Idea* (Judy Singer, 2017), based on her 1998 Honours thesis.

14. Thomas Armstrong, *The Power of Neurodiversity: Unleashing the Advantages of Your Differently Wired Brain* (Da Capo Lifelong Books, 2011).

15. www.england.nhs.uk/mental-health

16. Thomas Armstrong, *The Power of Neurodiversity: Unleashing the Advantages of Your Differently Wired Brain* (Da Capo Lifelong Books, 2011).

17. The Union of the Physically Impaired Against Segregation and the Disability Alliance, 'Fundamental Principles of Disability', summary of discussion held on 22 November 1975.

18. Michael Oliver, *Social Work with Disabled People* (Macmillan Education, 1983).

19. Patrick Campbell, Christopher Constantino and Sam Simpson (eds), *Stammering Pride and Prejudice: Difference Not Defect* (J & R Press, 2019).

8. Virtuous Disfluency

1. Sarah Bradford, *George VI: The Dutiful King* (Penguin, 2011).

2. Aneurin Bevan, *In Place of Fear* (William Heinemann, 1952).

3. Nicklaus Thomas-Symonds, *Nye: The Political Life of Aneurin Bevan* (I.B. Tauris, 2014).

4. Michael Foot, *Aneurin Bevan: A Biography* (Scribner, 1974).

5. John Mather, MD, 'Churchill's Speech Impediment Was Stuttering', article for International Churchill Society (2002).

6. Winston Churchill, 'The Scaffolding of Rhetoric' (unpublished manuscript, c.1897).

7. Vernon Bogdanor, 'Aneurin Bevan and the Socialist Ideal' (Gresham College lecture, 2012).

8. 'Putting it Bluntly', *W Magazine* (1 October 2007).

9. Elizabeth Inchbald, *A Simple Story* (London, 1791).

10. Chrissy Iley, 'Dan Ackroyd: a comedy legend's spiritual side', *Telegraph* (28 February 2012).

11. Richard Laliberte, 'Actor Dash Mihok on How Tourette Syndrome Shaped His Career', *Brain and Life* (October 2019).

12. Renée Byrne and Louise Wright, *Stammering: Advice for All Ages* (Sheldon Press, 2008).

13. John Hendrickson, 'What Joe Biden Can't Bring Himself To Say', *The Atlantic* (Jan/Feb 2020).

14. Mona Greenfield and Ellayne S. Ganzfried, *The Word Escapes Me: Voices of Aphasia* (Balboa Press, 2016).

15. Ibid.

16. 'Stammering and Identity: Land of Too Much' (producer Jayne Egerton), BBC Radio 4 (15 May 2013).

17. Steven Connor, *Beyond Words: Sobs, Hums, Stutters and Other Vocalizations* (Reaktion Books, 2014).

18. David Mitchell, 'Let Me Speak', *Telegraph* (30 April 2006).

19. Marc Shell, *Stutter* (Harvard University Press, 2005).

20. David Shields, *Dead Languages* (Alfred A. Knopf, 1989).

21. C.D. Dye, M. Walenski, S.H. Mostofsky, M.T. Ullman, 'A Verbal Strength in Children with Tourette Syndrome? Evidence from a Non-Word Repetition Task', *Brain and Language* (September 2016).

22. Fanny Burney, *Journals and Letters* (Penguin Classics, 2001).

23. Jason W. Brown (ed.), *Jargonaphasia* (Academic Press, 2013).

24. David Crystal, *How Language Works: How Babies Babble, Words Change Meaning and Languages Live or Die* (Penguin, 2007).

25. Steven Pinker, *The Language Instinct* (William Morrow and Company, 1994).

26. Darcey Steinke, 'My Stutter Made Me a Better Writer, *New York Times* (6 June 2019).

27. Jonathan Bryan, *Eye Can Write: A Memoir of a Child's Silent Soul Emerging* (Lagom, 2018).

9. The Art of Disorder

1. Lewis Carroll, *The Complete Works of Lewis Carroll* (The Nonesuch Library, 1939).

2. Morton N. Cohen, *Lewis Carroll: A Biography* (Vintage Books, 1995).

3. Ibid. The original quotation comes from G.J. Cowley-Brown, 'Personal Recollections of the Author of "Alice in Wonderland"', Scottish *Guardian* (28 January 1898).

4. Ibid. The original quotation comes from H.T. Stretton, 'More Recollections of Lewis Carroll – II', *The Listener* (6 February 1958).

5. Collected in Stuart Dodgson Collingwood, *The Life and Letters of Lewis Carroll* (Thomas Nelson and Sons, 1898).

6. Morton N. Cohen, *Lewis Carroll: A Biography* (Vintage Books, 1995).

7. Caryl Hargreaves, 'Alice's Recollections of Carrollian Days, as Told to her Son', *Cornhill Magazine* (July 1932).

8. Morton N. Cohen, *Lewis Carroll: A Biography* (Vintage Books, 1995).

9. Lewis Carroll, 'Alice on the Stage', *The Theatre* (April 1887).

10. J. de Keyser, 'The Stuttering of Lewis Carroll', *Neurolinguistic Approaches to Stuttering* (Brussels, 1972).

11. Morton N. Cohen (ed.), *The Selected Letters of Lewis Carroll* (Macmillan Press, 1982).

12. Ibid.

13. Darcey Steinke, 'My Stutter Made Me a Better Writer', *New York Times* (6 June 2019).

14. Christy Brown, *My Left Foot* (Secker & Warburg, 1954).

15. Mona Greenfield and Ellayne S. Ganzfried, *The Word Escapes Me: Voices of Aphasia* (Balboa Press, 2016).

16. Euan Ferguson interview with Edwyn Collins, '"I couldn't really talk. The words I could say were 'yes', 'no' and 'the possibilities are endless'"', *Guardian* (28 September 2014).

17. Grace Maxwell, *Falling and Laughing: The Restoration of Edwyn Collins* (Ebury Press, 2010).

18. Edwyn Collins, *Understated* (AED Records, 2013) and *Balbea* (AED Records, 2019).

19. Marc Shell, *Stutter* (Harvard University Press, 2005).

20. Charles Dickens, 'Psellism', *Household Words* (November 1856).

21. Ted Morgan, *Somerset Maugham* (Jonathan Cape, 1980).

22. Somerset Maugham, *The Moon and Sixpence* (William Heinemann, 1919).

23. B.B. King, *Blues All Around Me: The Autobiography of B.B. King* (Avon Books, 1996).

24. John Lee Hooker, 'Stuttering Blues', *Don't Turn Me From Your Door* (Atco Records, 1963).

25. Charles Shaar Murray, *Boogie Man: The Adventures of John Lee Hooker in the American Twentieth Century* (St Martin's Griffin, 1999).

26. Van Morrison, 'Cyprus Avenue', *It's Too Late to Stop Now* (Warner Bros, 1974).

27. Laura Barton, 'A Duel with Van Morrison: "Is this a psychiatric examination? It sounds like one"', *Guardian* (31 October 2019).

28. Led Zeppelin, 'Whole Lotta Love', *How the West Was Won* (Atlantic Records, 2003).

29. Dan Reilly, 'Kendrick Lamar Reveals Childhood Stutter', *Spin* (26 June 2014).

30. Ed Sheeran's speech on stuttering, recorded in *Time* magazine (10 June 2015).

31. Susannah Gora, 'How Carly Simon Overcame Stuttering and Migraine', *Brain and Life* (November 2009).

32. John Updike, *Self-Consciousness: Memoirs* (Penguin, 1990).

33. James Boswell, *The Life of Samuel Johnson* (London, 1791).

34. Samuel Johnson, *A Dictionary of the English Language* (London, 1755).

35. Leon Edel, *Henry James: A Life* (HarperCollins, 1985).

36. Grace Maxwell, *Falling and Laughing: The Restoration of Edwyn Collins* (Ebury Press, 2010).

37. Samuel Beckett, *The Letters of Samuel Beckett: Volume 1, 1929–1940* (Cambridge University Press, 2009).

38. Laura Salisbury and Chris Code, 'Jackson's Parrot: Samuel Beckett, Aphasic Speech Automatisms, and Psychosomatic Language', *Journal of Medical Humanities* (June 2016).

39. Samuel Beckett, *Collected Poems* (Faber & Faber, 2013).

40. Oliver Sacks, 'Tourette's Syndrome and Creativity', *British Medical Journal* (December 1992).

41. Steph Harmon, 'Backstage in Biscuit Land Review', *Guardian* (18 October 2016).

42. Letter from Charles Dodgson to Henry Rivers, 19 December 1873: 'Just now I am in a bad way for speaking, and a good deal discouraged. I actually so entirely broke down, twice lately, over a hard "C", that I had to spell the word! Once was in a shop, which made it more annoying.' In a following letter (27 December 1873), he thanks Rivers 'for advice about hard "C", which I acknowledge as my vanquisher in single-hand combat, at present'.

10. Speech Acts of Resistance

1. Siegfried Sassoon, 'They', from *The Old Huntsman, and Other Poems* (William Heinemann, 1917).

2. Siegfried Sassoon, 'The General', from *Counter-Attack, and Other Poems* (William Heinemann, 1918).

3. Siegfried Sassoon, *Memoirs of an Infantry Officer* (Faber & Faber, 1930).

4. Robert Graves, *Good-Bye to All That* (Anchor, 1929).

5. Hugo Ball, *The Dada Manifesto* (July 1916).

6. Anna Lawton (ed.), *Russian Futurism Through its Manifestos, 1912–1928* (Cornell University Press, 1988).

7. Peter Leese, *Shell Shock: Traumatic Neurosis and the British Soldiers of the First World War* (Palgrave Macmillan, 2002).

8. Geoffrey O'Hara, 'K-K-K-Katy' sheet music (Leo Feist, 1918).

9. Siegfried Sassoon, 'Survivors', from *Counter-Attack, and Other Poems* (William Heinemann, 1918).

10. Quoted in Chris Eagle, *Dysfluencies: On Speech Disorders in Modern Literature* (Bloomsbury, 2013).
11. Peter Leese, *Shell Shock: Traumatic Neurosis and the British Soldiers of the First World War* (Palgrave Macmillan, 2002).
12. R.S. Norman, C.A. Jaramillo, B.C. Eapen, M.E. Amuan, M.J. Pugh, 'Acquired Stuttering in Veterans of the Wars in Iraq and Afghanistan', *Military Medicine* (April 2018).
13. W.H.R. Rivers, *Instinct and the Unconscious* (British Psychological Society, 1919).
14. Siegfried Sassoon, *Sherston's Progress* (Faber & Faber, 1936).
15. Quoted in Ben Shephard, *Headhunters: The Pioneers of Neuroscience* (Vintage Books, 2014).
16. L.S. Jacyna, *Lost Words: Narratives of Language and the Brain, 1825–1926* (Princeton University Press, 2000).
17. Henry Head, *Aphasia and Kindred Disorders of Speech,* (Cambridge University Press, 1926).
18. Gilles Deleuze and Felix Guattari, *Capitalism and Schizophrenia*, Volume 1, Volume 2, *Anti-Oedipus*, and *A Thousand Plateaus* (Paris: Les Editions de Minuit, 1972 and 1980).
19. Dambudzo Marechera, 'An Interview with Himself', from *The House of Hunger* (Heinemann, 2009 edition).
20. Ibid.
21. Ibid.
22. Doris Lessing's review of Marechera's *The House of Hunger* in *Books and Bookmen* (June 1979).
23. Donna Ferguson, 'Greta Effect Leads to Boom in Children's Environmental Books', *Guardian* (11 August 2019).
24. Greta Thunberg, TEDxStockholm talk (24 November 2018).
25. Christian Laes, 'Silent History? Speech Impairment in Roman Antiquity', in Christian Laes, C.F. Goodey and M. Lynne Rose (eds), *Disabilities in Roman Antiquity* (BRILL, 2013).
26. Peter Dominiczak, 'Ed Balls: Reaction to My Stutter "really upsetting"', *Telegraph* (22 October 2013).

27. Hélène Mulholland, 'Ed Balls: "I won't apologise for my stammer"', *Guardian* (6 December 2012).

28. For biographical details of Wittgenstein's life I have drawn predominantly from Ray Monk, *Ludwig Wittgenstein: The Duty of Genius* (The Free Press, 1990).

29. Ludwig Wittgenstein, *Lectures: 1930–1932* (Rowman and Littlefield, 1980).

30. Ibid.

31. Lawrence Goldstein, *Clear and Queer Thinking: Wittgenstein's Development and his Relevance to Modern Thought* (Duckworth, 1999).

32. Julian Bell, 'An Epistle on the Subject of the Ethical and Aesthetic Beliefs of Herr Ludwig Wittgenstein', *The Venture* (February 1930).

33. G. Kreisel, 'Critical Notice: "Lectures on the Foundations of Mathematics"', in S.G. Shanker (ed.), *Ludwig Wittgenstein: Critical Assessments* (Croom Helm, 1986).

34. Norman Malcolm, *Ludwig Wittgenstein: A Memoir* (Oxford University Press, 1958).

35. Quoted in Ray Monk, *Ludwig Wittgenstein: The Duty of Genius* (The Free Press, 1990).

36. Rudolf Carnap, 'Autobiography', in Paul Schlipp (ed.), *The Philosophy of Rudolf Carnap* (Open Court, 1963).

37. Ludwig Wittgenstein, 'Lecture on Ethics', *Philosophical Review* (January 1965).

38. Ludwig Wittgenstein, *Culture and Value* (Wiley-Blackwell, 1998).

39. Alison Rourke, 'Greta Thunberg Responds to Asperger's Critics: "It's a superpower"', *Guardian* (2 September 2019).

40. David Shields, *Dead Languages* (Alfred A. Knopf, 1989).

11. Communication Diversity

1. Steven Pinker, *The Language Instinct* (William Morrow and Company, 1994).

2. Ibid.

Notes

3. Edouard Glissant, *Caribbean Discourse: Selected Essays* (University of Virginia Press, 1992).

4. Aimé Césaire, *Notebook of a Return to My Native Land* (Bloodaxe, 1995).

5. John Patrick Walsh, *Free and French in the Caribbean: Toussaint Louverture, Aimé Césaire, and Narratives of Loyal Opposition* (Indiana University Press, 2013).

6. Frantz Fanon, *Black Skin, White Masks* (Grove Press, 1967).

7. Jean Bernabé, Patrick Chamoiseau, Raphaël Confiant and Mohamed B. Taleh Khyar, 'In Praise of Creoleness', *Callaloo* (Autumn, 1990).

8. Alexander Graham Bell, 'The Utility of Signs in the Instruction of the Deaf', *The Educator* (1898).

9. Andrew Solomon, *Far From the Tree* (Vintage, 2014).

10. William Stokoe, *Sign Language Structure: An Outline of the Visual Communication Systems of the American Deaf* (University of Buffalo, 1960).

11. Andrew Solomon, *Far from the Tree* (Vintage, 2014).

12. Ibid.

13. For example, Doreen Lenz Holte, *Voice Unearthed: Hope, Help and a Wake-Up Call for the Parents of Children Who Stutter* (Holte, 2011).

14. For instance, as of 2016, the Employers Stammering Network (ESN), launched in 2013 to raise awareness about disability rights and best practice, claims to represent an employed population of 1.5 million. The ESN inspired, in turn, the UK Civil Service Stammering Network and the Defence Stammering Network.

15. Francesca Martinez, *What the **** is Normal?!* (Virgin Books, 2015).

16. Lee Ridley, aka Lost Voice Guy, *I'm only in it for the Parking: Life and Laughter from the Priority Seats* (Bantam Press, 2019).

17. Vyvyan Evans, *The Emoji Code: How Smiley Faces, Love Hearts and Thumbs Up are Changing the Way We Communicate* (Michael O'Mara, 2017).

18. Anna Doble, 'UK's Fastest Growing Language is … Emoji', *BBC News* (19 May 2015).

Epilogue: Out of the Mouth Trap
1. Grace Maxwell, *Falling and Laughing: The Restoration of Edwyn Collins* (Ebury Press, 2010).

Acknowledgements

I am grateful, first and foremost, to those who guided me, made crucial introductions and shared personal experiences. They include: Marina Abramović, Giuliano Argenziano, Nihal Arthanayake, Sebastian Barfield, Mary Beard, Jamie Beddard, Patrick Campbell, Brian Catling, Alison Clark, Oliver Dimsdale, Margaret Drabble, Will Eaves, Max Egremont, Daniel Everett, Robert Douglas Fairhurst, Tim Fell, Hartry Field, James Fox, Betony Kelly, Danny Ladwa and the Stammering Voice Orchestra, Charlotte Mosley, Hans Ulrich Obrist, Scroobius Pip, Janina Ramirez, Rebecca Roache, Uri Schneider, Sophie Scott, Walter Scott, Owen Sheers, Rory Sheridan, David Shields, Jim Smith, Debris Stevenson, Colm Tóibín, Jennifer Vonholstein, Kate Watkins, Tom Wheeler and Harry Yeff.

Ngũgĩ wa Thiong'o, David Shields, Joshua St Pierre and Jess Thom were interviewees who significantly shaped my ideas at critical stages in the writing process. Roland Allen, Dan Fox, Henry Hitchings and Thomas Karshan are writers and friends whose advice I sought and was thankful for on a number of occasions.

While there are many books about speech disorders, there are few on their cultural and historical context. For this reason, I relied on a handful of important exceptions: *Knotted Tongues* by Benson Bobrick, *A Cursing Brain?* by Howard I. Kushner, *Dysfluencies* by Chris Eagle, *Lost Words* by L.S. Jacyna and *Stutter* by Marc Shell.

I have been guided or supported by Elaine Kelman, Frances Cook and Jo Hunter from the Michael Palin Centre for Stammering Children, by Carolyn Cheasman from City Lit, and by Tim Fell and Jane

Powell from the British Stammering Association. The achievements of all three organisations are nothing short of heroic. I am also extremely grateful for the support of Jane Fraser and the Stuttering Foundation of America and to Professor Shelagh Brumfitt who read the book at draft stage. But most of all, I thank therapist Willie Botterill, whose expertise and care I have depended upon at various, often vulnerable stages in my life.

As a needy first-time author, I suspect I have occasionally been a drain on the time and patience of my publishers. The idea for the book was developed in close collaboration with the brilliant Kirty Topiwala at the Wellcome Collection. Andrew Franklin provided invaluable advice and necessary interventions throughout. Ellen Davies is a superb and diligent editor who oversaw the final drafts. The wider teams at Wellcome and Profile have been incredibly supportive, including Penny Daniel, Hannah Ross and Joe Staines.

My greatest thanks go to my family. My parents, Annie and Stephen, and sister, Tamsin, were always supportive when I felt most lost in speech and language. My mother has always encouraged me to write, although probably never expected to find herself copy-editing the last draft of her forty-four-year-old son's book. My cousin Gilly was generous in sharing her own difficulties with words. She is a beacon of kindness and bravery; an inspiration to all who know her. My wife's family, the Wyndhams, all have personal experience, to varying degrees, of stuttering and shared memories and ideas as the book progressed.

And, of course, the endeavour would have been impossible if not for my wife, the wonderful Constance Wyndham, who continued to encourage me even when the reality of life with two small children made the thought of our writing a shopping list feel a stretch too far, let alone a book.

Index

Note: *italic* entries are the titles of books, films, plays or performances discussed.

Index

Index

Index

Index

politicians with speech problems
Aneurin Bevan 15, 202–6,
224; Ed Balls 269, 292; Joe
Biden 88–9, 206, 209; Winston
Churchill 49, 204–5
politics, fluency prejudice 106–7
The Power of Neurodiversity
184–5, 188
premonitory urges 33, 166
presentation literacy 103–4, 109,
112
*The Presentation of Self in
Everyday Life* 86
problematic words and sounds
26, 63, 77, 83, 213, 232, 252
professions *see* career choices
psellism / psellismologists 18, 26,
127
*Psycho-Analysis and the War
Neuroses* 138
*Psycho-analytical Observation on
Tic* 140
psychoanalysis
and autism 178; decline 144–6,
149, 179; interpretation of
speech disorders 139–43, 145,
156; legacy 150–1, 156–7, 169;
peak influence 56, 142, 175,
178; shell shock and influence
136–8, 259–61; and Tourette's
syndrome 139–41, 143, 145,
156
psychological impact 51–2, 68,
71–3, 206, 211, 223
psychology, emergence 123–4,
134
*The Psychopathology of
Everyday Life* 136–7
psychotherapy, disappointing
impact 18
PTSD (post-traumatic stress
disorder) 137–8, 144, 256–60

public reading, fear of 232
public speaking, fear of 14, 84,
96, 104–5, 132, 205–6, 208, 227
public speaking techniques 101–
4, 108, 303–4

Q

quality of life improvement 10,
167
*Quiet: The Power of Introverts
in a World That Can't Stop
Talking* 105

R

rap music 147, 151, 215, 242, 292,
294
Reality Hunger 248, 249
Reith, John 2
relationships and speech
disorders 304
repetitions, part-word, word and
phrase 97–8
research funding 44, 72
rhetoric 112–13, 115–16, 118–19,
204, 298
Ridley, Lee (Lost Voice Guy) 93,
172, 187, 291–2
Rivers, W. H. R. 133, 259–61
Robinson, Ken 103
Robinson, Nick (broadcaster) 42,
170, 292
rock music 241–2
Rodman, George 110
Royal College of Speech and
Language Therapists 9, 286
Rustin, Lena 152

S

Sacks, Oliver
godfather of neurodiversity
175–8, 183, 190; identifies
ebullient qualities 177, 200,